Protecting Danube River Basin Resources

Ensuring Access to Water Quality Data and Information

NATO ASI Series

Advanced Science Institutes Series

A Series presenting the results of activities sponsored by the NATO Science Committee, which aims at the dissemination of advanced scientific and technological knowledge, with a view to strengthening links between scientific communities.

The Series is published by an international board of publishers in conjunction with the NATO Scientific Affairs Division

A Life Sciences **B Physics**	Plenum Publishing Corporation London and New York
C Mathematical and Physical Sciences **D Behavioural and Social Sciences** **E Applied Sciences**	Kluwer Academic Publishers Dordrecht, Boston and London
F Computer and Systems Sciences **G Ecological Sciences** **H Cell Biology** **I Global Environmental Change**	Springer-Verlag Berlin, Heidelberg, New York, London, Paris and Tokyo

PARTNERSHIP SUB-SERIES

1. Disarmament Technologies	Kluwer Academic Publishers
2. Environment	Springer-Verlag / Kluwer Academic Publishers
3. High Technology	Kluwer Academic Publishers
4. Science and Technology Policy	Kluwer Academic Publishers
5. Computer Networking	Kluwer Academic Publishers

The Partnership Sub-Series incorporates activities undertaken in collaboration with NATO's Cooperation Partners, the countries of the CIS and Central and Eastern Europe, in Priority Areas of concern to those countries.

NATO-PCO-DATA BASE

The electronic index to the NATO ASI Series provides full bibliographical references (with keywords and/or abstracts) to more than 50000 contributions from international scientists published in all sections of the NATO ASI Series.
Access to the NATO-PCO-DATA BASE is possible in two ways:

– via online FILE 128 (NATO-PCO-DATA BASE) hosted by ESRIN, Via Galileo Galilei, I-00044 Frascati, Italy.

– via CD-ROM "NATO-PCO-DATA BASE" with user-friendly retrieval software in English, French and German (© WTV GmbH and DATAWARE Technologies Inc. 1989).

The CD-ROM can be ordered through any member of the Board of Publishers or through NATO-PCO, Overijse, Belgium.

Series 2: Environment – Vol. 24

Protecting Danube River Basin Resources

Ensuring Access to Water Quality Data and Information

edited by

Irene Lyons Murphy
Colorado State University,
Fort Collins,
Colorado, U.S.A.

Springer-Science+Business Media, B.V.

Proceedings of the NATO Advanced Research Workshop on
An International Data-Sharmge Programme for the Effective Management
of Danube River Basin Resources
Budapest, Hungary 27–30 May 1996

A C.I.P. Catalogue record for this book is available from the Library of Congress

ISBN 978-90-481-4785-4 ISBN 978-94-017-2805-8 (eBook)
DOI 10.1007/978-94-017-2805-8

Printed on acid-free paper

CONTENTS

PREFACE

The Danube River Basin is a unique example of a collective response to a set of serious regional environmental problems. Immediately following the end of the Cold War western and eastern countries in the Basin, encouraged and guided by international agencies, sought to achieve a unified approach to upgrade the protection and management of vital natural resources. The NATO Workshop brought together key specialists to make recommendations for an essential feature of this goal: the sharing of accurate data and information about the water quality of the river and its tributaries.

The Workshop participants, many of whom have dual roles as scientists and government officials, provided papers, responded to questionnaires, collaborated across country borders on case studies of hot spots in transboundary Danube tributaries, and then gave four days of their time to intensive discussions of a key river basin question. They deliberated and came up with constructive answers to the primary question posed by the NATO Advanced Research Workshop, An International Data-Sharing Programme for the Effective Management of Danube River Basin Resources: How can the large and steadily increasing volume of data and information about water quality be shared among all Danube countries to their greatest mutual advantage?

This book is testimony to the contribution of all 24 workshop participants and to three who were unable to be present but whose papers were read and had an impact on ARW results. The support from the NATO Science Programme came at a crucial time, as the ad hoc agreement supporting regional improvements in the protection of Danube resources was about to be transformed into a legally binding convention. NATO Programme reports on other aspects of environmental security were helpful to developing the rationale for the meeting. Its funding rules accommodated the subject matter well and contributed to a successful workshop.

The co-director of the Workshop, Dr. Péter Bakonyi, and members of his staff at VITUKI Consult Rt in Budapest, worked with speed and efficiency on matters mundane and intellectually challenging. Dr. Bakonyi's long experience in the technical and policy sides of water management in the region was essential to the success of the Workshop. The Workshop was co-sponsored by and received strong support from the Department of Environmental Sciences and Policy at the Central European University. Its director, Dr. Edward Bellinger, who has an extensive background in UK and Danube water policies, provided guidance and generally facilitated workshop proceedings.

A debt of gratitude is especially owed to the other members of the organising committee. The extensive background of Professor Evan Vlachos, Colorado State University, in international river basin management helped with the development of all aspects of the Workshop. Dr. Atanas Paskalev, Bulgaria, is a practising engineer whose interests extend to public participation in environmental protection and promoting greater interest in the scientific aspects of water resource management. Dr. Aurel Varduca, Romania, has played a leading role in the first Basin-wide water quality data-sharing program which began before the transition period under the 1985 Bucharest Declaration.

The accommodations provided by the residence of the Central European University in Budapest, Kerepesi, were excellent in all respects. Its staff, headed by Katalin

Sellyei, responded promptly and efficiently to all requests, including supply of complex electronic equipment. Note-takers from the Central European University, Janice Forry and Bill Hanlon, United States, and Lidija Jasnic, Croatia, worked at a fast pace to transcribe the proceedings and provide complete and accurate notes of the sessions for this editor in record time. Ann Sherrington, working at VITUKI Consult Rt, guided the preparation of the papers presented at the Workshop.

Above all, this book commemorates the dedication of workshop participants to regional co-operation. Their contributions were grounded in skills developed as they met the difficult and tedious challenges of international collaboration in the early transition years. Let us hope that the state-of-the-art approach to sharing data and information they recommend may become a permanent part of peaceful co-operation to protect the resources of the Danube River Basin.

Irene Lyons Murphy

THE WORKSHOP:
Content, format, conclusions

IRENE LYONS MURPHY
Colorado State University
2005 37th Street NW
Washington, DC, 20007USA

PÉTER BAKONYI
VITUKI Consult Rt.
Kvassay Jenő út 1.
Budapest, Hungary

1. A Framework for Workshop Objectives

The NATO Advanced Research Workshop which met in Budapest from the 27th to the 30th of May 1996 had a specific objective: to recommend an action strategy for the optimal sharing of water quality data among Danube Basin countries. Its ultimate goal: to contribute to the effective management of water resources shared by 90 million people.

Panel sessions followed a progression from the general, the nature of ecological data sharing and its support through international agreements, to the specific, the optimal strategy for water quality data-sharing in the Danube Basin. The group of experts engaged in lively discussion from the start. Discussion is reflected in changes the authors have made in their original papers as well as in addenda to papers.

One immediate result of the first day's discussion was a change in the workshop title. It might be assumed that "data-sharing" also included "information sharing." But this should not be taken for granted. Information interprets, focuses data. For basin management objectives an optimal strategy has to include both.

To formulate the action strategy the workshop brought together 24 scientists and policy specialists from the Danube region, including Austria, Germany and eight countries in transition, and from the NATO countries of the Netherlands, the United Kingdom and the United States. Of the ten Danube countries represented, one, Germany, is a NATO member; eight are NATO partner countries.

After considering the eco-system needs of the Basin these specialists, who came from international agencies, government ministries, and research institutes, contributed to discussions of current environmental monitoring in other international river basins, water quality monitoring systems in four NATO countries and in the Basin countries in transition. They reviewed the work of the five-year Environmental Programme for the Danube River Basin and the needs of the soon-to-be permanent

I. Lyons Murphy (ed.), Protecting Danube River Basin Resources, 1–15.

secretariat of the International Commission for the Protection of the Danube River. They watched demonstrations of information and data systems available on diskette and the Internet. And then they spent the final day and a half poring over and revising a proposed action strategy.

2. Need for an action strategy

Participants had agreed to commit to the preparation of an action strategy for the development and sharing of accurate and compatible water quality data (add information) among Danube River Basin countries. The adopted strategy was built on the achievements of the Environmental Programme for the Danube River Basin (EPDRB) and the contributions of river basin specialists in the Basin and in other river systems. It was to be scientifically and technically sound and to be sure of the support of basin countries. Duplication of activities already under way by the EPDRB and/or the Secretariat of the International Commission for the Protection of the Danube River (International Commission) clearly was to be avoided. Cost effectiveness was a prime consideration.

As specialists and policy makers, workshop members were aware of the long history of transboundary water resource conflicts and the fact that some have been resolved and others left to sour relationships between basin countries. For centuries riparian countries in navigable international rivers have reached, and kept, agreements to maintain and improve the use of shared river resources for trade and commerce. The need to regulate flow, preserve channel depth, remove obstructions and assure international equality in customs and transport fees was recognized for centuries in the Danube Basin and in transboundary rivers throughout the world. Riparian countries committed to internationally conceived and enforced regulations despite historic conflicts which divided them in other ways. Navigation agreements required, among other commitments, the sharing of accurate, timely hydrologic data. Damage to ships and loss of trade had to be avoided if nations were to prosper.

Competition over water use comes first in a priority list of the issues which must be resolved if peaceful and equitable basin management is to be achieved. It is, therefore, necessary to go beyond acceptance of navigation regulation and include in negotiation settlements significant other issues related to resource use. Increasingly, diversion of river resources for domestic and industrial use, in particular the development of hydroelectric power projects, has become an issue between countries. Use of international rivers and their tributaries as disposal sites for municipal and industrial waste and the failure of countries to treat such waste have made water quality top items on the management agenda.

Principles of international law, more honored in the breach than in the observance, apply to river systems, notably the concept that each basin country should have a reasonable and equitable share in the beneficial use of shared water resources. Such sharing has been widely construed to incur an obligation to exchange hydrologic and other relevant data and information on a regular basis.

Commitment to equitable use of river basin resources must begin and be sustained by a commitment to share relevant data and information freely and without reservation.

3. Coordination on the Danube

Participants in the NATO Advanced Research Workshop recognized the work accomplished to date on applying these principles to the Danube River Basin. All basin countries needed to commit to an international convention which would recognize the need to share information, and to adopt a strategy for resource protection, including international investment policies.

3.1. TRANSITION OPPORTUNITIES

The political changes at the end of 1980s provided an opportunity to broaden and strengthen basinwide cooperation. The advantages to Western and Eastern Europe were clear. The west wanted to ensure that the environmental policies adopted under European Union council directives in the 1970s and '80s would be acceptable in the east. The new governments in Eastern Europe recognized the need to meet such standards if they were to trade with the west, become members of the European Union, and receive investment help for lagging economies. The European Union PHARE programme (EU PHARE), the Global Environmental Facility (GEF), the United States Agency for International Development (US AID) and other funding organizations sought and received cooperation from all of the basin countries in a joint move to improve basinwide management.

In the fall of 1991 Danube basin countries and the international funders reached an ad hoc agreement which set in motion a series of significant steps toward improved environmental policies. A task force of representatives of international agencies, country governments, and environmental organizations established a Danube Environment Programme (DEP) and agreed on a detailed work plan in early 1992. The plan provided for national reviews of the status of environmental policies in each country, the development and timely sharing of compatible data, commitment to a strategic action plan to establish priorities for cleanup projects, and the harmonization of legislative environmental standards and their enforcement. The DEP and the countries allied with it have worked hard and successfully together to fulfill this plan.

3.2. CRITICAL TIME FOR DANUBE COOPERATION

The action strategy which resulted from this workshop comes at a critical time. The DEP has reached the end of its first phase which established a groundwork for a long-term management programme, one which emphasized environmental issues but inevitably related to all phases of the economic and political transition of post-1989 Eastern Europe. Work under the direction of EPDRB has proven the need for the free exchange of all data and information relevant to the goals of sound environmental management. It has substantiated the relevance of such data and information and moved significantly toward optimal methods of sharing it, based on state-of-the-art technology. An international action strategy is needed which will capitalize on past and current investments in improved water quality data, guided by knowledge and experience in a number of scientific and technical disciplines and feasible in terms of cost and international and national goals.

This is also the time for funding and other decisions about the future of the EPDRB and the implementation of the Danube River Protection Convention. The goals of this workshop have attracted world-wide interest. Aside from their obvious importance to the river basin itself they support as well those of the NATO Science Programme on Environmental Security, the resolution of regional environmental problems and the adoption of "affordable cleanup technologies." River basin commissions elsewhere in Europe and throughout the world should gain from the contributions of this workshop. Workshop deliberations and conclusions should receive attention from all those concerned about the rationalization of transboundary management of environmental conflicts.

4. Agreement to Share Knowledge: A Sign of Trust

4.1. KEYNOTING A PROGRESSION

Sir Richard Southwood, a distinguished zoologist and former vice-chancellor of Oxford University, provided a keynote for the workshop framework [DATA AND DATA-SHARING: *Key to Transboundary Water Quality Protection*, page 17]. The ultimate objective served by data and information about the Danube River "eco-system," however interesting they may be in themselves, must be to support the better and sustainable management of the natural resources. The materials requiring monitoring are identified and summarised. They are part of a continuity of the river eco-system which must be understood to achieve efficient management.

The attitudinal and technical aspects of data-sharing need to be addressed. Harmonisation of monitoring methods, for example, requires continued international attention with regard especially to scientifically complex areas. The goal of sustainable development, advanced by the Brundlant Commission report and endorsed by the UN Conference in Rio de Janeiro in 1992, applies in all respects to the Danube. "The Danube Basin has had more than its share of conflict. Sharing knowledge is a sign of trust and leads to understanding: this is an important step on the road to mutual support and sustainable development."

4.2. THE IMPORTANCE OF AGREEMENTS

4.2.1 *Lessons From The Rhine Agreement*

From the panel on international agreements we sought to know more about what cooperation on the Rhine had to offer a workshop on the Danube. The Rhine Basin is close geographically, but very different in political and economic development. Formal agreements to monitor and share data pre-dated those on the Danube by several decades. It faces management challenges similar in some respects and yet very different in others. Like river basins everywhere its improved management resulted from firm agreements to assess water quality and other data, and to set goals on this basis for needed tributary and river cleanup.

Dr. Marc Braun [THE RHINE: *Monitoring programmes and exchange of data within the icpr*, page 23] has been a member of the ICPR since 1988. He has specialised in measurement programmes (water, suspended matter), analytical quality assurance, quality targets, and best available techniques for industrial sectors.

He reminds us that the Rhine is one of the most intensively monitored international rivers in the world. It has had an international convention since 1976 which requires its members (Switzerland, France, Germany, Luxembourg and the Netherlands) to install and operate international monitoring stations and to finance them. It publishes monitoring results annually and has had an action programme since 1987 which sets targets for the reduction of 50 priority substances for established uses. Co-operation, while it has been effective, is not compulsory since the Rhine Action Plan (RAP) is not a convention binding under international law but rather a "gentlemen's agreement." All measures adopted under the RAP are voluntarily implemented by the states bordering the Rhine. "The principle of a maximum of transparency is applicable to the implementation of measures, and applies to the exchange of data and the results of the monitoring programmes and inventories."

In the target programme concentration has been fixed for each of the priority substances, taking into account each asset meriting protection. Such integrated management based on an ecosystem approach requires large-scale as well as medium- and long-term observations of the quality status of the Rhine, which suggests major goals.

These goals can only be reached by applying sensitive, quantitative, standardised and reliable monitoring methods which have proved effective in analytical practice. There are differences in size and other factors as Dr. Braun has pointed out between the Rhine and Danube. Rhine countries have conformed to European norms and thus have shared compatible monitoring and data-sharing methodologies. The Sandoz disaster of 1986 argued strongly against complacency and triggered a more definitive data-sharing programme. The need for state-of-the-art monitoring and data-sharing is as great in each basin and to a considerable extent must be met in similar. if not the same, ways.

4.2.2. *The First Danube International Environmental Agreement*

Dr. Aurel Varduca, who has co-ordinated the monitoring activities set in motion by the 1985 Bucharest Declaration since 1991, has a degree in analytical chemistry and is the head of the water quality monitoring and database division of the environmental research and Engineering Institute, Bucharest, Romania [THE 1985 BUCHAREST DECLARATION: *An Important First Step in International Danube Water Quality Monitoring*, page 31].

During 1985-1987 a common monitoring programme was established under the Bucharest Declaration by agreement among Austria, Bulgaria, the former Czechoslovakia, Germany, the former Yugoslavia, Romania, Hungary and the former Soviet Union. Thirteen stations began to collect samples at Danube sites for agreed upon parameters as well as to record the rate of flow on a monthly basis. These data have been shared once a year since 1988.

General objectives included overall ecological protection for Basin resources. Need for the more recent concepts for a transnational monitoring network, and international data comparison was recognised. A two-year (1988 to 1990) transboundary monitoring programme was followed by a second phase which improved methodology and extended the range of determinands. Analytical methods follow those of the American Standard Testing Material of the American Water Works Association. A special file for exchange of information on emergency situations has been maintained.

Urgent needs, most of which are being met by the present monitoring programme activities of the EPDRB, include improved network design, selection of additional monitoring parameters, special analyses of other river basins with respect to the economic costs of monitoring, and more intensive capacity building for technology, including laboratory services. Dr. Varduca describes many other equally urgent needs for improved data-sharing in the Danube Basin.

In spite of significant improvements in monitoring systems after 1985, through the use of a common set of parameters and sampling methods, the resulting data bases available were not satisfactory for decision makers especially for the lower Danube River. He believes that the improvements now generated by the EPRDB monitoring sub-group are constrained by many gaps in our knowledge concerning the functioning (including fluctuations along the three spatial dimensions of Danube River System), productivity and carrying capacity of most of the ecosystem types which can be identified within the river system. These need to be explored over time to continue support for optimal monitoring and data-sharing.

4.2.3. A Danube River Protection Convention

Wilhelm Kittinger heads the section on international water management co-operation and supranational water management (relating to the European Union) for the Austrian Federal Ministry of Agriculture and Forestry. He also serves as President of the interim International Commission for the Protection of the Danube River (International Commission). His discussion of the history of the Danube River Protection Convention (DRPC) [THE DANUBE RIVER PROTECTION CONVENTION: Provisions for Environmental Data-Sharing, page 43] stresses the co-operative spirit of the negotiations which preceded the adoption of the Convention in 1994. Signed by nine Basin countries it has been formally ratified by five countries and is expected to go into force by 1996-1997. The atmosphere of good neighbourliness was aided by co-operative experiences under the Bucharest Declaration and the Convention on the Protection and Use of Transboundary Watercourses and International Lakes (Helsinki Convention) which was signed at Helsinki on 17 March 1992.

The International Commission has closely co-operated with the EPDRB. The latter's Strategic Action Plan provides that the responsibility for the water-related parts of the EPDRB be transferred to the International Commission once the DRPC's entry into force is realised.

The scope of the DRPC, which applies to the entire catchment area of the Danube River, includes all water-related activities and issues. Some core provisions are specified in detail, others are part of a framework which will be further developed by the International Commission. In addition, this framework will provide for the International Commission's systems for internal data sharing as well as for a data bank and public information access. Such networks will take into consideration all aspects of information exchange.

Kittinger cautions that constraints on human and financial resources as well as the difficult economic conditions and the domestic budgetary problems for most of the contracting parties call for efficiency and the avoidance of duplication. This applies both to the implementation of the DRPC as well as to the responsibilities to be transferred from the EPDRB to the International Commission as far as water-related issues are concerned.

5. Data-sharing Programmes and Policies

The NATO countries represented on the panel shared the same water quality data challenges and had developed similar technology to deal with them. Germany, the Netherlands, and the United Kingdom, as members of the European Union, have met EU Council directive goals with respect to water quality standards and their enforcement. Germany and the Netherlands are members of the International Commission for the Protection of the Rhine and adhere to its Action Programme mandates. The United States has effected an assessment system which responds to similar needs but faced unique challenges in terms of the complexity of its river basin systems. The workshop panel and discussion focused on specific problems in the development of monitoring and data-sharing programmes in these four western democracies. By summarising issues and accomplishments they sought to provide a framework for the panelists from the East European and CIS Danube countries for possible changes in their own environmental data-sharing policies.

5.1. GERMAN MONITORING AND DATA-SHARING PROGRAMMES

Prof. Dr. Hans Liebscher, who serves as a senior specialist with the Federal Office of Water in Germany, notes the effects of massive emissions of anthropogenic contaminants into the atmosphere, into soil, and into waters which increased during post-World War II development in both agricultural and urban areas [GERMAN MONITORING AND DATA-SHARING PROGRAMMES, page 49]

Negative effects on all water uses created a need to monitor pollution loads and trends. By providing a basis for political decision-making, monitoring programmes have contributed to reductions of classical pollution loads such as heavy metals and chlorinated hydrocarbons in most German rivers and coastal waters.

By the mid-1950s Germany had begun to establish monitoring programmes for inland watercourses and marine waters. Since the mid-1960s the Federal States have intensified the installation of new water quality monitoring stations. New knowledge about the pollution of waters and advances in measuring technology have contributed to the enormous expansion of water quality monitoring in recent times.

Extensive monitoring systems exist but Dr. Liebscher considers them to be adequate and cost efficient only in part. Monitoring programmes which include chemical-analytical methods have produced a confusingly vast amount of data which gives little information about the actual state of the environment. Only by integrating the above-mentioned approaches with simulation models and by combining them with geographic information systems (GIS) and with remote sensing can these disadvantages be avoided and the costs of monitoring reduced. Such an approach has already been considered in some of the new ecosystem approaches.

5.2. WATER QUALITY PROGRAMMES IN THE NETHERLANDS

Dr. Harm Niederländer is a researcher with the International Centre for Water Studies in Amsterdam. He is an analytical chemist by profession who has worked on many aspects of monitoring and assessment programmes. His detailed review of water quality monitoring in the Netherlands [DUTCH (INTER)NATIONAL WATER QUALITY DATA PROGRAMMES, page 65] stresses the support given to the definition and attainment of intermediate and ultimate water quality targets by monitoring programmes. The results of water quality assessment and the assessment of discharges to surface water are periodically published in a number of reports in the Netherlands and made widely available.

International commitments have had an impact. The Rhine Action Programme required the reduction of emission loads of priority substances within a predetermined period of time, i.e. from 1985 to 1995. Data collection through inventory and monitoring is required to assess the effectiveness of policy approaches. In addition, ambient water quality assessment programs reporting on an annual basis are being considered by monitoring working parties, operating under the International Commission for Protection of the Rhine against Pollution and the Oslo and Paris Commissions.

5.3. MONITORING AND DATA-SHARING IN THE UK

Dr. Edward Bellinger, a water quality specialist from the United Kingdom who presently heads the Department of Environmental Sciences and Policy at the Central European University in Budapest, emphasises the long history in the UK of changes in the regulatory framework of water pollution control that have resulted from frequent modifications by successive governments [WATER MANAGEMENT MONITORING IN THE UNITED KINGDOM, page 81]. In 1989 the largely public water supply and regulatory bodies were replaced by a privatised industry and a separate, independent regulatory authority. A new regulatory organisation, the Environmental Agency, was formed in March of 1996. These and other changes are explored with reference to monitoring and data exchange.

The Environmental Protection Act of 1990 (Part I) introduced some further changes to water pollution control by making the most dangerous discharges (Red List Substances) subject to prior approval under a system of Integrated Pollution Control (IPC). Embodied in all of these provisions is a requirement to monitor and make information available. The information may not necessarily be published but should be accessible in some form to the public.

Dr. Bellinger concludes that there is a general need to monitor water courses for their quality in order to ascertain what and how improvements may be made so that the resource may be enhanced. The information gathered must be made freely available to all interested parties so that meaningful discussions and participation can take place in working for overall improvements. In order to achieve these improvements some mechanism for policing any agreed standards must exist and it must have teeth enough to work.

5.4. THE US WATER QUALITY ASSESSMENT PROGRAM

David Wangsness is a senior hydrologist with the United States Geological Survey's (USGS) National Water Quality Assessment Program [DESIGN AND IMPLEMENTATION OF THE NATIONAL WATER-QUALITY ASSESSMENT PROGRAMME: *A United States example*, page 89]. The Assessment, which became operative in 1991, reports status and trends in surface- and ground-water quality at national, regional, and local levels. The legislative basis for US monitoring and data-sharing policies are identified as well as the successive phases of the design and implementation of the USGS Assessment. With respect to the Danube, he believes that priorities and issues could produce a priority list of data and information needs. Based on a review of available data an assessment programme could be designed to fill the gaps, taking care to standardize approaches, techniques, and data bases, and to use new data in conjunction with existing data to address the priority questions. Standardization is critical throughout the programme.

5.5. COUNTRIES IN TRANSITION

A symposium provided an opportunity for the Danube countries in transition to discuss their water quality data-sharing policies. The countries participating were Croatia, Slovenia, Czech Republic, Slovak Republic, Hungary, Romania, Bulgaria, and Ukraine. The Moldovan representative sent a reply to the survey conducted for the Workshop but was unable to attend. Please see page 101 for a symposium summary by the editor.

6. Danube Basin Water Quality Data-Sharing

6.1. CURRENT DATA-SHARING ACTIVITIES

The status of current Danube River Basin water quality data-sharing programmes was reviewed in detail in a day-long panel session. One covered the present water quality data and information sharing which is under the guidance of or assisted by the EPDRB. It included presentations on the Programme itself and on the extensive work of its sub-group on Monitoring, Laboratory and Information Management (MLIM) as well as a discussion of public participation activities. A three-basin project on international basin tributaries funded by the United States Agency for International Development and the Global Environmental Facility was presented in a paper by the project director.

The second set of presentations dealt with the sharing of data at the field level in five case studies, presented by experts from the countries sharing Danube Basin tributaries and/or the river itself.

6.1.1. *The Environmental Programme for the Danube River Basin*

David Rodda was the team leader of the EPDRB's Programme Co-ordinating Unit from 1992 until March 1996. He is now a consultant appointed for the European Commission PHARE Programme and an adviser to the present team leader of the PCU, Teun Botterweg. He is a civil engineer by profession and has specialised in water environmental management and operation. He describes the scope and work of the Environmental Programme for the Danube River Basin and the provisions of the agreements which support the exchange of data and information [DEVELOPMENT OF INFORMATION ACTIVITIES IN SUPPORT OF THE ENVIRONMENTAL PROGRAMME FOR THE DANUBE RIVER BASIN, page 113]. Related achievements and activities, past and future, are discussed.

The case for and the possible details of a data and information strategy are examined, covering the marshalling and dissemination of existing documented information, the handling and exchange of measured or time-series data, and the development of tools which transform data into information and its dissemination. Mr. Rodda discusses the assignment of responsibility to the Danube countries for data and information dissemination, the advantages of network links between systems rather than designing and implementing new system(s) and the cost-benefits of such activity.

6.1.2. *The work of the monitoring sub-group*

Dr. Tim Lack, Water Research centre (WRc) UK, is also the Project Leader of the European Topic Centre on Inland Waters (European Environment Agency) [THE WORK OF THE MONITORING, LABORATORY AND INFORMATION MANAGEMENT SUB-GROUP, page 123] . He has worked as a principal consultant and project director in the field of pollution assessment and control and the design of monitoring programmes. He is the Project Manager of the team of consultants that proposed the Trans-National Monitoring Network (TNMN) for the Danube River Basin and leader of the team that has designed a water and quality monitoring network for the countries of the European Environment Agency.

The measures for establishing the Programme network include the identification of a TNMN of representative stations for the assessment of the chemical and biological quality of the Danube and its major tributaries and the selection of National Reference Laboratories for the Danube in each riparian country. An information management system for the TNMN supports exchange of data, statistical trend analysis and interfaces with national information systems in each Danube riparian country. Training is also provided for system operators. Rather than specify a data and information management system to be used by all countries, the protocol by which countries should interchange data from the TNMN has been defined. This Data Exchange File Format (DEFF) will interface with each National Information System; the resource requirements necessary have been identified.

6.1.3. *Support for Public Participation*

Robert Atkinson is on the staff of the Regional Environmental Center for Central and Eastern Europe (REC) where he serves as the REC officer responsible for the Danube Grants Programme [THE DANUBE GRANTS PROGRAMME AND THE NGO ROLE IN MANAGEMENT OF THE DANUBE BASIN, page 135]. Public participation in the development of policy on data-sharing and other aspects of Danube activities is supported by the Danube Grants Programme. It is managed by the Regional Environmental Center and funded through the Global Environment Facility. The grants programme evaluates how these independent activities assist the protection of the Danube Basin, particularly where they involve a monitoring or research component. The ceiling for grants to eligible nongovernmental organizations (NGO) is set at ECU 4,000 per year per organisation.

Other organisations working on environmental issues in the Danube River Basin are also eligible, such as scientific associations, farmers organisations, co-operatives, scientific and technical institutions, provided that they worked together with community-based NGOs or local community groups.

6.1.4. *Three Basin Project*

Ron Orach is the director of a Metcalf & Eddy International, Inc. project to detect and reduce transboundary pollution sources on Danube River tributaries and to minimise the effects of these sources on downstream water users [TRANSBOUNDARY POLLUTION REDUCTION IN THREE DANUBE RIVER BASINS, page 149]. The project will implement water quality improvements in Hungary, Slovakia, and Romania, providing equipment, technical assistance, and training in connection with a variety of discrete tasks in each country. These tasks are designed to either prevent/control sources of pollution or to mitigate the downstream impacts of pollutants. The project is one of implementation rather than study and includes a strong emphasis on the supply and installation of equipment.

6.2. FIVE CASE STUDIES IN DATA SHARING

Bi- and multi-lateral agreements to settle management issues on transboundary tributaries have existed for centuries in the Danube region. Some resolved boundary concerns, others suggested the regulation of navigation. In the post-World War II years, as water quality monitoring increased, so did an interest in sharing monitoring data. Workshop participants from Danube countries were asked in this panel to help determine the status of transboundary data-sharing. Five case studies are described, covering a range of tributaries and, in one case the exchange of data along the Danube River itself. Case study participants were invited to be in touch with each other prior to the workshop, and to submit a collaborative paper and/or be prepared to discuss the situation from the point of view of the country he or she represented.

Discussion focused on present data-sharing policies and recommendations for improving collaboration on the Inn River (Austria and Germany); Morava Basin (Czech Republic and Slovak Republic); the Drava (Slovenia, Austria, and Hungary); Upper Tisza (Hungary, Ukraine, Romania, Slovakia), and the reach of the Danube shared by Bulgaria and Romania.

6.2.1. *The Inn River*
Karl H. Schwaiger is the head of the Water Quality Monitoring and Assessment section of the Austrian Federal Ministry of Agriculture and Forestry specialising in the design, co-ordination and management of Austria's water quality monitoring network [AUSTRIAN-GERMAN DATA-SHARING. page 165]. He reported close cooperation between the two countries in water quality monitoring and data-sharing along the 650-kilometre stretch of the Inn which serves as a border between the two countries. About 150 kilometres of the border follow, in part, the course of the Rivers Inn, Salzach and Danube.

All transboundary rivers in this region flow northwards from the Austrian Alps and discharge into Bavaria. Considerably enlarged by the discharge from the Bavarian territory, their waters re-enter Austria via the Danube at Jochenstein. Germany and Austria are therefore simultaneously upstream as well as downstream countries from each other. He outlines the core parts of the legal basis for water quality monitoring as well as the range of practical experiences which have modified it. There are several bilateral Austrian-German projects worth noting such as conservation of wetlands, ground water modelling, among others. The focus is on the sharing of water quality data, the benefits of tailor-made monitoring and quality assurance programmes.

Dr. Liebscher reported from the German side that cooperation in monitoring and sharing data is satisfactory.

6.2.2. *The Morava*
Marcela Fabianová, Slovak Hydrometeorological Institute, Slovak Republic and Jozef Drozd, Morava River Basin Authority, Czech Republic, collaborated on the Morava case study presentation. Ms. Fabianová is currently the National Focal Point for the Danube Environmental Programme. Dr. Drozd is the assistant chief of water laboratories for Povodi Moravy (Morava River Basin Authority) in Brno, Czech Republic.

The paper provides an overview of activities related to surface water quality monitoring, primarily within the Morava River Basin, with special attention to exchange of data and information between the Slovak Republic and the Czech Republic [DATA-SHARING BETWEEN THE SLOVAK AND CZECH REPUBLICS IN THE MORAVA RIVER BASIN, page 173]. It includes basic information on the Morava River Basin, its monitoring network and international activities related with surface water quality.

The River Morava is one of the most important tributaries of the Danube River. The total area of its Basin is 26,658 km^2 most of which (approximately 84 %) belongs to the Czech Republic. The Morava River and its largest tributary, the Dyje. create a single hydrological unit. Their basins spread over three states: the Czech Republic, Austria and the Slovak Republic. The Morava River, which has a total length of 352 km, rises in the Czech Republic and flows into the Slovak Republic. The length of the border between the Slovak and the Czech Republics is approximately 30 km. The downstream Morava River creates a border between the Slovak Republic and Austria approximately 70 km in length up until its confluence with the Danube River.

It is clear that monitoring and quality control of surface waters is not and cannot be the problem of only one state; it is at least a bilateral issue. Transfer of information and data exchange between involved partners play a very essential role in their management. The authors cite the need to improve and maintain surface water quality within the Morava Basin through improved exchange of information between the Slovak Republic and the Czech Republic. Such information would help to identify the amount, type and source of pollution.

6.2.3. *The Upper Tisza River Basin*

Dr. Ferenc László, of the Water Resources Research Plc (VITUKI Rt) in Budapest, Hungary, contributed the paper about the Upper Tisza [SHARING DATA ACROSS NATIONAL BOUNDARIES IN THE UPPER TISZA RIVER BASIN, page 183]. Since he was unable to participate in the Workshop, it was read by Dr. Péter Literáthy, director of the Water Resources Center of the Institute for Water Pollution Control in Budapest who participated in the discussion. He was joined in the case study discussion by Valodomir Bilokin of Ukraine, and Corneliu Negelescu of Romania.

The Upper Tisza River Basin is shared by Ukraine, Romania, Slovakia and Hungary. It is the longest (977 km) tributary of the Danube and also has the largest drainage basin. The catchment area of the Upper Basin is 49,500 km^2. Bilateral transboundary water committees in the Upper Tisza Basin began in the 1960s and 1970s to organize regular joint water quality investigations and evaluations and to develop joint measures to respond to accidental pollution.

At present regular water quality investigations are carried out jointly with neighboring countries at transboundary stations; measurement results are evaluated annually by special transboundary committees and trend analyses are made available every fifth year. Accidental pollution spills which might have transboundary impacts are subject to special information-sharing requirements.

In addition to the bilateral exchange of data a preparatory study for investment, financed by the European Bank for Reconstruction and Development, has also contributed to the exchange of environmental data in the Upper Tisza Region. In the framework of this programme the consulting agency has collected, with the assistance of local organizations, the available environmental data of the entire Bodrog river catchment and of the Upper Tisza Basin upstream of the confluence with Bodrog.

Volodimir Bilokon, a member of the panel, is the head of the water resources department in the ministry for environmental protection and nuclear safety of Ukraine. During the discussion he cited the effectiveness of the Ukraine-Hungary commission in the development of joint protocols for sampling techniques. Supervision of the Tisza's water quality is conducted within the framework of this arrangement. The annual meetings of delegations permit the discussion of water quality problems which may have arisen in the previous year. An emergency warning system is scheduled to be established in Uzjgorod.

Corneliu Negelescu, Research and Engineering Institute for Environment, Bucharest, introduced additional data about the Tisza. The country's National Water Quality Monitoring System characterises the quality of Upper Tisza water as very good.

6.2.4. *The Drava River Basin*
D. Sc. Jasna Grbovič and D. Sc. Meta Gorišek of Slovenia participated in the discussion of the Drava Basin, presenting a paper [CROSS-BORDER MONITORING OF WATER QUALITY IN SLOVENIA, page 189] which covered responsibilities for cross-border monitoring of water quality in Slovenia as it applied to transboundary monitoring generally. Slovenia has agreements to settle water management issues on the Drava River with Austria and Hungary.

Since 1992 all data have been brought together in the Hydrometeorological Institute computer centre in an integrated data base. Monitoring of ambient waters in Slovenia has been performed since 1965 when the sampling points (locations) network was composed of 20 locations while today there are 110. Commissions made up of representatives from Austria and the former Yugoslavia were established for monitoring purposes in 1954 for the Drava river and in 1955 for the Mura. The availability of long-term monitoring data is considered to be very important for transboundary water quality issues. The sampling data of only a few years duration are insufficient for decision making and negotiations about the common goals and the performance of international water management.

6.2.5. *The Bulgarian-Romanian Reach of the Danube*
Dr. Atanas Paskalev is the Director of Water Resources and Ecology, for Cesi-Consult, Ltd, specialising in the computer aided design of water supply and sewerage systems. He has completed a number of projects in water supply and waste water treatment in Bulgaria and is the author of several articles on these subjects [THE BULGARIAN-ROMANIAN REACH OF THE DANUBE: *Suggestions for improving monitoring and data-sharing,* page 199]. The boundary between Bulgaria and Romania in the lower part of the Danube River suggests the importance of creating and supporting harmonised and mutually relevant laws and regulations for the protection of the water quality of the rivers discharging into the Danube River from each country.

Toxic substances pollute surface and groundwater in the Danube region through waste water discharges, surface runoff, atmospheric fallout and leaching from solid waste disposal sites. In addition to these chronic pollution inputs the water quality along the river is influenced by accidental pollution, man-made structures and natural processes. Consequently the pollution from the Danube catchment to the Black Sea is responsible for eutrophication in the north-western sea areas.

Dr. Paskalev looks critically at the progress of implementation of recent monitoring agreements. The enforcement of each agreement, as well as data and information exchange, are closely related to the harmonisation of member countries' legislation. The need for environmental legislation and standards harmonisation is especially emphasised in the Danube Convention in Article 9 and in the Strategic Action Plan for the Danube River Basin. Both countries seek admission to the EU and must live up to its criteria for environmental management. The key components to a realistic monitoring and enforcement programme are outlined.

Dr. Aurel Varduca, Ministry of Waters, Forestry and Environmental Protection-Environmental Research and Engineering Institute, Bucharest, Romania presented a brief statement concerning data quality monitoring on the Danube. He cited as advantages to increased data-sharing between the two countries monitoring under the

Bucharest Declaration of 1985 and the 1991 Convention between the Governments of Romania and Bulgaria on cooperation in the protection of the environment as well as the anticipated increase in data-sharing which should take place under the Danube River Protection Convention and the current work of the MLIM for the EPDRB.

7. Developing Consensus on Strategy

The sessions which preceded the adoption of the Conclusions and Recommendations raised serious concerns about water quality data and information-sharing in the Danube Basin. Questions about the control of both on the Internet, the optimal format for information, and how to proceed in the most efficient and cost-effective way were resolved, at least for now.

DATA AND DATA-SHARING:
Key to transboundary water quality protection

SIR RICHARD SOUTHWOOD
Zoology Department, University of Oxford
South Parks Road
Oxford, UK

KEYWORDS / ABSTRACT : Danube River Basin / water quality monitoring / water quality data sharing

A river basin must be viewed as a dynamic system governed by underlying flows, both physical and biological. The interrelationships of the processes are determined by their position in the longitudinal gradients of the catchment from headwaters to estuary. Industrial and urban developments impose changes on natural processes in this continuum, not necessarily at the point where they occur but throughout the course of the river. Information is needed on physical and biological components, as well as their properties such as the speed and turbidity of the water. A classification schema is suggested to relate materials carried by the water to their ecological roles. Pollution control, its process, costs and benefits are summarised. Information sharing is a sign of trust and leads to understanding; hence the importance of the NATO Workshop.

1. Introduction

The Danube has many claims to be described as Europe's greatest river. It is the longest, rising in the Black Mountains and entering the Black Sea after flowing for nearly 2900 km. It rises further west than Denmark and it ends further east than some parts of Asia. It virtually travels the breadth of our continent, passing through or adjacent to nine countries - and more are served by its tributaries. It is a great natural signal of our interdependence one on another - we are all linked together by this great artery as are the parts of the body.

I will not attempt to address any of the details of the workshop topic in this brief introduction. I will start by giving an ecologist's view of a river basin, then I will consider some general issues that arise in data sharing. Although information has an inherent interest, it is the better and sustainable management of this natural resource that is the ultimate objective. When conclusions are drawn from the data, therefore, these are likely to indicate that some current practices must change. How should these be viewed, as burdens or opportunities?

I. Lyons Murphy (ed.), Protecting Danube River Basin Resources, 17–21.

2. The river system as an ecosystem

The Danube is a great ecosystem; the interplay and functioning of its physical and biological components refuse to recognise any man-made, political boundaries. But human activities modify this system as a whole, and the flow from headwaters to sea gives a strong directional bias to the influence of any actions. This bias is of course the basis for the simplest form of pollution control - having one's outflow upstream from one's intake!

A river basin may be described as a physical geomorphic system dominated by water, whose flow from headwaters to mouth normally presents a gradient of physical conditions - velocity, volume, temperature and materials being carried. Turbidity often, but not always, follows the same gradient.

The materials carried by the water that will need to be monitored may be classified in various ways in relation to their ecological role (e.g. Southwood & Conway, 1975 [1]). Basically we may recognise:

1) **Materials required for life** - dissolved oxygen and nutrients. Oxygen is captured from the air more particularly as the water moves quickly and in a thin stream, as over falls and weirs or across the rocky bed of a shallow stream. Oxygen is also contributed to the water by the photosynthesis of submerged water plants and algae - but this is rate-dependent on the sunlight penetrating the water. Oxygen is removed by all organisms as they respire and, if there are large numbers of non-photosynthetic organisms (e.g. bacteria breaking down organic material), there may be a shortage of oxygen, and fish, etc. will die. Key data are therefore the oxygen level and the BOD (Biological Oxygen Demand). Nutrients are normally measured in terms of the relevant ions - nitrates, phosphates, etc. If the levels of nitrates and/or phosphates are excessive eutrophication occurs, there is an overproduction of algae and similar organisms, excessive oxygen demand, and widespread deaths in fish and other biota.

2) **Waste products, natural or man-made that may be utilisable**. Carbon dioxide from the respiration of organisms is a universal waste product, but it is of course a vital raw material for photosynthesis. Many other wastes, such as sewage or paper pulp waste, may be used by some parts of the community, but whether they are detrimental or not is summarised by the advertising slogan "not too little, not too much": there is an optimum. Pollution is not an absolute characteristic, it depends on concentration/time [2].

3) **Materials that may be detrimental to life**. For example pollutants deposited from the air, sulphate ions in acid rain, or lead from the emissions of vehicles burning petrol with lead additives, to give two man-made examples. Phenols are an example of waste products that are both natural and man-made in origin and always detrimental to life. Recently there has been some concern that hormones excreted by women taking contraceptive pills may enter the river and eventually the public water supply, with detrimental effects on the male population.

4) **Materials entering the system by leaching or erosion**. These may be directly harmful, as when the ground waters pass through rocks containing toxic materials (e.g.

copper, zinc), or like most soil particles arising by erosion having indirect effects (in this case causing turbidity), or even beneficial if they provide a source of nutrients.

5) **Organisms** carried in the water, from fish of large size and high commercial value - the sturgeon of caviar fame - to the viruses that cause a variety of gastro-enteritic and other disorders and whose significance was hardly appreciated a few decades ago.

Although water cannot be destroyed or created, only the minority of water that falls as rain or snow in the catchment area of a river will eventually flow through its estuary to the sea. For the Rhine, for example, this has been calculated as about one-third. Some is retained in ice (depending on the climatic trend), some in porous rock and so to underground aquifers, much evaporates, either directly (sometimes "aided" by man e.g. large open water surfaces or agricultural sprayers operating in wind and sunshine) or through plants - transpiration. Lastly, after extraction by man some is returned as effluent to another river system. However, so far as the ecosystem is concerned, there is a continuity in the water and the recognition of this led to the development of the "River Continuum Concept" by Vannote *et al.* (1980) [3].

This concept recognises the gradient of biological communities reflecting the physico-geochemical gradients in the water from headwaters to sea. Communities will evolve to make the most efficient use of the energy in the sector they occupy - energy sources being sunlight, the materials carried in the water, and other organisms. Seasonal fluctuations in physical conditions constrain the exactness with which this efficiency can be achieved; this leads to leakages (unutilized energy), but leakages upstream provide additional resources downstream. Thus an understanding, and ultimately the efficient management, of the river system depends on a knowledge of the values of the physico-geochemical parameters in all sections - no section is isolated or totally independent.

3. Attitudinal and technical aspects of data sharing

3.1. POLITICS OF DISPUTES OVER TRANSBOUNDARY EFFECTS

My own experience in the dispute between the Scandinavian countries and Britain over the significance of British flue gas emissions in relation to the acidification of surface waters in southern Scandinavia (4) showed me how in such a situation the protagonists tend to emphasise the uncertainty of the scientific evidence. If the scientists can be adequately isolated from national pressure, then progress can be achieved if the governments or others concerned can be persuaded in advance to agree to accept the conclusions of a multinational scientific team.

3.2. EXTERNALITIES AND THE PERCEPTION OF THE COSTS OF SUSTAINABLE MANAGEMENT

In the simplest analyses the polluter is seen to pass material to the receiver (the polluted). The former receives all the benefits, the costs are externalised and fall very largely on the receiver. The application of "the polluter pays" principle is often seen to simply transfer the cost back to the polluters, to the relief and benefit of the receivers. As I will show, pollution abatement may bring economic benefit to the polluter such that one has a win/win situation.

3.3. HARMONISATION OF MONITORING METHODS

The comparison of different measurements is only justified if they are rigorously standardised; this applies to the criteria for sampling just as much as to the measurements themselves. Laboratories have many approaches to achieve quality control but it not an aspersion on any system to insist on the exchange of samples between laboratories. In scientifically complex areas, e.g. the measurement of aluminium species, there can be a considerable advantage in holding international workshops addressed to the particular issue.

4. Objectives

The objective is laid down in the title of the workshop, "the effective management" of the river basin. Today we may be sure that this means the sustainable development of the Danube basin. The concept of sustainable development was advanced in the Brundtland Commission's report "Our Common Future" and endorsed by the UN Conference in Rio de Janeiro in 1992. We must therefore be concerned about changing the nature of the basin by pollution and about utilising water so as to, in some way, adversely affect its availability in the future.

4.1. POLLUTION CONTROL - THE PROCESS, COSTS AND BENEFITS

Sharing data will enable the problems to be recognised. Improvements may be made in the short or long term. Short-term improvements generally involve the use of the existing plant with some modifications in working practices. Several major international manufacturing companies run schemes in which the whole of their workforce is invited to put forward proposals. These normally relate to a particular plant, and the various plants compete for the honour of having their innovation judged as being worthy of an award.

In the pharmaceutical company, Glaxo Wellcome plc, where I am a non-executive Director and chair the award-making panel, there are nearly 40 entries each year. This process has three advantages: the whole workforce is involved, the changes are normally inexpensive (they may indeed save money), and the cumulative effect of many such changes is considerable. The management has to address major problems; if action is urgent this will often require plant modification, retrofitting, this is always expensive. The most economic way in which to bring about significant changes in a manufacturing process is to incorporate the necessary modifications in the specifications for new plant. This means that environmental policy needs to be long-term, objectives should be set and adhered to, recognising in the case of manufacturing processes that the most cost effective timescale will correspond to that of the re-equipment cycle.

When waste is reduced it is often found that this represents a significant cost saving. Although waste often appears to have no value to the person who wishes to dispose of it, in many instances it represents a waste of a resource. A good example is often provided by the recovery and reuse of organic solvents that can bring great savings.

As well as these direct benefits from pollution control, there are secondary effects; they are not simply a case of being a good neighbour, they are likely to have advantages in "image improvement." There are now a number of international standards for environmental management, such as ISO 14001 and the EU's EMAS, which signal adherence to appropriate standards. Purchasers are demanding that their suppliers hold such certification, which is likely to become an important feature of EU-wide trade.

4.2. WATER USE AND AVAILABILITY

Much use of water in Europe may be described as profligate rather than prudent - no doubt a reflection of the abundant rainfall that most of us have benefited from hitherto. For example, a significant proportion of the water that is sprayed overhead onto fields is lost by evaporation on hot and windy days. Consideration also needs to be given to storing more of the water that we receive during periods of rainfall; ground water reservoirs may be artificially enhanced.

5. Conclusions

As several of the papers to be presented indicate, much has already been done towards achieving your objectives. The Danube Basin has had more than its share of conflict. Sharing knowledge is a sign of trust and leads to understanding: this is an important step on the road to mutual support and sustainable development.

References

1. Southwood, T.R.E. and Conway, G.R. (1975) Man's effect on the ecology of the Rhine basin, in *Regional Management of the Rhine*. Papers of a Chatham House Study Group, European Series No. 26, pp. 11-14.
2. Southwood, T.R.E. (1995) Ecological processes and sustainability. *International Journal of Sustainable Development and World Ecology* **2**, 1-11.
3. Vannote, R.L., Minshall, G.W., Cummins, K.W., Sedell, J.R. and Cushing, C.E. (1980) The river continuum concept, *Canadian Journal of Fisheries and Aquatic Science* 37, 130-137.
4. Mason, B.J. (1992) Acid Rain: *The Acidification of Surface Water*, Cambridge University Press, Cambridge.

THE RHINE:
Monitoring programmes and exchange of data within the ICPR

Dr. M. BRAUN
International Commission for the Protection of the Rhine
Technisch-wissenschaftliches Sekretariat der IKSR
Postfach 309
D-56003 Koblenz

KEYWORDS / ABSTRACT: Rhine River Basin / water quality data-sharing / International Commission for the Protection of the Rhine / Rhine Action Programme

The Rhine catchment area provides drinking water for 20 million people and serves industrial purposes as well; it is one of the most intensively monitored international rivers. Monitoring strategies respond to emergency needs such as sudden inputs of pollutants as well as to record medium- and long-term changes in Rhine water quality. The International Commission on the Protection of the Rhine (ICPR) against Chemical Pollution (1976) obliges the member states (Switzerland, France, Germany, Luxembourg and the Netherlands) to install and operate international monitoring stations, including their financing. The Rhine Action Programme (RAP) (1987) set targets for the reduction of 50 priority substances for established uses.

1. The Rhine

From a hydrological point of view the River Rhine is not very impressive. Contrary to the Danube, which is 2,860 kms long and drains an area of 817,000 km^2, the Rhine is only 1,320 km long and drains an area of 185,000 km^2. With its long-standing mean discharge of 2,300 m^3/s the Rhine achieves only one third of the discharge rate of the River Danube.

The Rhine, however, is one of the world's busiest inland waterways. Rotterdam is the largest sea port of the world, Duisburg the largest inland port. Nine countries belong to the Rhine drainage area, five of them representing the main part of the drainage area with its 50 million inhabitants. The Basel-Rotterdam axis represents the core of the West European industrial boom which followed World War II. No other river system has as many chemical factories on its banks as the Rhine. Rhine water is being intensively used by the power-producing industry, by other types of industry and by agriculture. Rhine waterworks supply some 20 million people with drinking water.

Apart from the International Commission for the Protection of the Rhine (ICPR) another 20 international organisations co-operate within the catchment area of the

I. Lyons Murphy (ed.), Protecting Danube River Basin Resources, 23–29.

Rhine. These include the Central Commission for Navigation on the Rhine and the International Union of Waterworks on the Rhine, the last one representing a strong lobby for good water quality.

2. ICPR genesis and history

2.1. THE ICPR, CREATION AND EARLY DECADES

Population growth, economic growth and, above all, progressive industrialization led to an increasing degradation of Rhine water quality. The post-war economic boom did the rest.

Demand for concrete action grew louder, particularly from the Dutch, for whom a protected Rhine is of vital importance and who therefore urged the cooperation of all countries bordering the Rhine at an early stage of the intensive post-war development. In order to stop the continuous degradation of Rhine water quality, the Dutch initiated closer co-operation of riparian states soon after World War II. In 1950 Switzerland, France, Germany, Luxembourg and the Netherlands joined in the ICPR, thus creating a common body for negotiations on problems and possible solutions along the Rhine.

The Bern Convention of 1963, a convention binding under international law, first listed the obligations and tasks of the ICPR:

- To establish the extent and origin of pollution
- To develop measures aiming at protection of the Rhine
- To prepare corresponding international treaties.

From 1972 on the work of the ICPR has been complemented by Ministers' Conferences. During the extremely dry years of 1971/72 Rhine water quality reached its lowest point. Nothing was left of the traditional abundance of fish. People called the Rhine "the most romantic sewer of Europe."

2.2. CONVENTION TARGETS

In 1976 the ICPR drafted the Convention on the Protection of the Rhine against Chemical Pollution (Chemical Convention) which was ratified by the Member States. The same year, the European Community signed the Bern Convention as the sixth contracting party. The following targets were to be reached under the Chemical Convention:

- Reduction of inputs of hazardous substances by fixing threshold values for single substances (black list) and drafting of national programmes with a view to reducing less hazardous substances (grey list)
- Establishment of a riverwide alert system in case of sudden accidental pollutions of the Rhine
- Establishment of international monitoring stations and monitoring programmes for Rhine water.

From the mid 1970s to the mid 1980s a large number of wastewater treatment plants were built along the banks of the Rhine, so that, by the end of the 1980s, Rhine water quality had improved considerably.

The working methods of the ICPR and policies of the Rhine border countries changed over night as a consequence of the warehouse fire at Sandoz near Basel, which greatly affected the Rhine ecosystem.

3. The Rhine Action Programme

The Sandoz accident and its consequences made evident that sources of danger were not yet under complete control and demonstrated the extreme vulnerability of the Rhine ecosystem. Therefore, the Rhine border countries not only introduced measures aimed at limiting and remedying damages but also drafted an extensive cleanup programme with targets and deadlines which was adopted by the Ministers in 1987: The Rhine Action Programme (RAP).

By the year 2000 the implementation of the RAP is supposed to lead to the following achievements:

- A higher level of species, previously indigenous in the Rhine (e.g. salmon), is supposed to return to the river. This target represents the overall improvement to be achieved not only with respect to water quality, but to the entire ecosystem.
- The use of Rhine water for drinking water supplies must continue to be possible in future.
- The river sediment is to be treated to such an extent that it may be used for land filling or be dumped into the sea.
- In 1989 the Ministers added a further target: the improvement of the ecological state of the North Sea.

These targets are to be reached by implementing the following measures:

- Reduction of direct discharges (industry, municipalities) and of indirect inputs (atmosphere, agriculture)
- Reduction of accidental spills due to higher safety standards in industry
- Hydrological and morphological improvement of habitats for flora and fauna of the Rhine and its alluvial areas.

The RAP is not a binding Convention under international law but rather a "gentlemen's agreement." All measures adopted under the RAP are voluntarily implemented by the states bordering the Rhine. The principle of a maximum of transparency is applicable to the implementation of measures, and applies to the exchange of data and the results of the monitoring programmes and inventories.

4. Monitoring programmes, exchange of data and publications

A list of priority substances, of particular importance for the Rhine catchment area, has been drawn up within the RAP. Within a ten-year period, from 1985 to 1995, discharges of these substances were supposed to be reduced by 50%, in some cases even by 70%. The application of state-of-the-art-technology in industry and municipal wastewater treatment plants is supposed to reduce emissions, that of the best environmental practice is supposed to reduce diffuse inputs. Emission reductions of dischargers are monitored by regular inventories of discharges of priority substances and by a minimum surveillance programme. Additionally, the RAP also provides for a reduction of accidental spills by assuring sustainably increased safety standards. This double approach to emissions is being completed by quality targets set for about 50 priority substances. The approach combining aspects of discharges and of environmental quality permits the drafting of a comprehensive monitoring strategy thus obtaining a complete set of data on the basis of which it is possible to judge the quality of the Rhine.

4.1. TARGET VALUES

In order to be able to make a uniform assessment of the quality status of the Rhine, the ICPR has fixed target values. These are not threshold values, but standards for an ecotoxicological assessment of the Rhine. They emphasize assets meriting protection such as drinking water supply, aquatic communities, fishery and use of sediments. The quality of suspended matter, sediments and water are taken into account. A target concentration has been fixed for each of the priority substances, taking into account each of the assets meriting protection. The lowest of these values has been fixed as the ICPR target value for the Rhine. Eutrophication criteria are applicable to phosphorous.

4.2. MONITORING PROGRAMMES CONCERNING RHINE WATER QUALITY

Two different monitoring strategies have developed in the Rhine drainage area: the monitoring programme carried out under the alert system and the annual and five-yearly monitoring programmes on the medium- and long-term development of Rhine water quality.

4.2.1. *Early warning monitoring*

Under the alert system monitoring programmes are being carried out with two goals: the detection of sudden discharges and their reasons, the evaluation of the potential danger of the discharge and the rapid transmission of information about them. The first aim requires rapid and often semi-quantitative methods of analysis, recording a broad spectrum of substances. Monitoring results must be available within a few hours, so that measures like stopping the water intake from the Rhine or retaining contaminated water in collection basins may be implemented in a timely manner. Fully automatic monitoring systems, such as on-line analytics, are recommended. Specifically adapted data processing, long-distance data transfer and automatic sampling systems provide for 24 hour-a-day monitoring.

Reports are passed on downstream to national and local warning centres according to predetermined schemes. The Convention on Chemical Pollution obliges the Member States to pass on any information on accidents threatening the Rhine water quality to other Member States as well as to the ICPR-Secretariat. Local, not international, organisations are in charge of taking adequate steps in case of accidental spills. Since local organisations are equally in charge of discovering and assessing sudden discharges, an international harmonisation of analytical methods as well as of monitoring programmes is not necessary.

4.2.2. *International programme monitoring the quality of the Rhine water*
According to the Convention on Chemical Pollution the Member States are obliged to run monitoring stations and equivalent monitoring systems. Monitoring results must be reported to the ICPR at least once a year. Apart from international monitoring stations there are a number of national and regional ones. Monitoring programmes of the national monitoring stations are co-ordinated among one another and with the international monitoring programme. According to the Convention on Chemical Pollution the ICPR is obliged to draft recommendations for monitoring programmes and methods of analysis and is in charge of the comparability of analysis results.

Measurement results are entered into ICPR data-masks on discs by the operators of the monitoring stations, interpreted by a central computer and examined by the secretariat with respect to their plausibility. The results are published annually and are available as printed matter or on computer disc. At this time, the monitoring programme comprises about 90 parameters in water and 40 in suspended matter.

Since the RAP has been adopted, the annual monitoring programme has been completed by an extensive five-yearly monitoring programme, assessing the following compartments:

- chronology and longitudinal profiles of contents of noxious substances in water, suspended matter and sediments
- contamination of Rhine fish
- species diversity and population density of fish, macrozoobenthos and plankton.

5. Specification of the targets of the monitoring programmes

The most important step when developing a successful monitoring programme at low costs is to clearly define the targets and the information needed. The targets and accordingly the required information result from the RAP. This integrated management based on an ecosystem approach requires large-scale as well as medium- and long-term observations of the quality status of the Rhine, which suggests the following goals:

- Comparison of the quality of water and suspended matter with the target values
- Long-term development of the concentrations of noxious substances in water and suspended matter and of the loads in the Rhine.

These goals can only be reached by applying sensitive, quantitative, standardised and reliable monitoring methods which have proved effective in analytical practice. The monitoring systems must respond to higher demands regarding accuracy and reliability than those used for the international alert system. Since ICPR measures regarding the reduction of point source and diffuse inputs are based on the target values, this particularly applies to the comparison of the quality of water and suspended matter with the target values. Frequency of sampling, time of sampling, sampling method as well as parameters monitored are subject to international agreement so as to grant comparability of the results of analysis in the longitudinal profile of the Rhine.

The ICPR provides that the targets of the international monitoring programme are realised with minimum costs and a maximum of comparability and information. Within the monitoring programme the methods, frequency and the substances to be monitored are fixed annually one year ahead of monitoring. The organisation in charge of the monitoring station is responsible for the implementation and financing of the monitoring programme.

6. Examples of the use of data obtained by the monitoring programmes and inventories

Since the construction of the wastewater treatment plants in the mid-1970s concentrations of BOD_5 have continuously decreased. Even before the Sandoz accident occurred, oxygen saturation reached 90% at the German-Dutch border. Additionally, an inventory of municipal and industrial BOD_5-discharges revealed that about 95% of all inhabitants of the Rhine catchment area discharge their polluted water into wastewater treatment plants and thus monitoring programmes regarding BOD_5 and COD were stopped.

Since the 1980s the substitution of phosphorous in detergents and its precipitation in wastewater treatment plants has resulted in a marked decrease of total phosphorous and phosphorous loads in the Rhine. It is estimated that the target value of 150 mg/l had been reached in 1996.

Contrary to the experience with phosphorous compounds, nitrate nitrogen has increased on a long term basis. Its loads in the Rhine are partly due to leaching from farmland. The target of a 50% reduction between 1985 and 1995 will not even be reached by the year 2000, as measures aimed at reducing diffuse inputs are only effective on a long term basis.

Concentrations of heavy metals in water are today below limit values for drinking water. Since heavy metals accumulate in suspended matter and sediments and recycling of sediments were a decisive factor when fixing the target values, a monitoring programme concerning suspended matter was added to the international monitoring programme in 1990. The annual comparison of Rhine water quality with the target values reveals that the target values fixed for these substances have not been reached at all monitoring stations. The results of the last comprehensive 5-yearly comparison show that in a certain number of fish the heavy metal contents were in excess of threshold values for food.

The inventory of priority substances revealed that the major part of AOX and chloroform discharges originated from the paper pulp industry. Therefore, the state-of-the-art-technology was first fixed for this industrial branch. AOX-concentrations measured at the Koblenz/Rhine monitoring station show that AOX loads have been significantly reduced.

Contrary to most other substances the concentrations of HCB in sediments and suspended matter decline in the longitudinal profile of the Rhine. A comparison of HCB emissions and their contents in sediments showed that the major part of the HCB-concentrations in the Rhine are due to former discharges having settled in the barrage weirs of the Upper Rhine.

Continued analyses and trend data support short- and long-term policy decisions and have provided, overall, a satisfactory way of sharing both water quality data and information in the Rhine Basin.

THE 1985 BUCHAREST DECLARATION:
An important step toward danube water quality protection

AUREL VARDUCA Ph.D.
Research and Engineering Institute for Environment
Bucharest, Romania

KEYWORDS / ABSTRACT : Danube River Basin / Bucharest Declaration / water quality monitoring

The Danube River Basin is an aquatic ecosystem of high economic, social and environmental value. It provides drinking water supplies, agriculture, industry, fishing, tourism and recreation, power generation and navigation for a densely populated region of the European continent. The region contains many important natural areas, including wetlands and flood-plain forests. During 1985-1987 a common monitoring programme was established and agreed to under the Bucharest Declaration by Austria, Bulgaria, the former Czechoslovakia, Germany, the former Yugoslavia, Romania, Hungary and the former Soviet Union. Thirteen stations began to collect samples at transboundary sites for agreed upon parameters as well as to record the rate of flow on a monthly basis.

These data have been shared once a year since 1988. The Research and Engineering Institute for Environment which has co-ordinated the Bucharest Declaration monitoring programme since its inception sees urgent needs for expanded monitoring programmes in the Basin and is co-operating with the EPDRB and the International Danube Commission to achieve such goals.

1. The Danube River and its Basin

The Danube River has always been one of Europe's most important waterways, providing basic necessities for everyday life such as food, water and energy as well as a transportation and a communication link. The Danube River Basin is no less than the heartland of Central Europe, while the Danube Delta, where the river discharges to the Black Sea, is the second longest natural wetland area in the continent. Main tributaries include the Tisza, Sava and Drava, as well as the Inn, Morava and Prut.

The waters of the Danube River Basin and its tributaries comprise an aquatic ecosystem of high economic, social and environmental value. They support drinking water supplies, agriculture, industry, fishing, tourism and recreation, power generation and navigation for a densely populated region of the European continent. The region contains many important natural areas, including wetlands and flood plain forests. The

I. Lyons Murphy (ed.), Protecting Danube River Basin Resources, 31–41.

Danube River Basin is an area of great ecological richness, biodiversity, within Europe. It provides habitats for over 100 species of fish out of the total of 227 found in Europe. In the Danube Delta alone, 300 species of birds are found. In 1985 an important agreement to monitor pollution in the river, embodied in the Bucharest Declaration began the first international monitoring system in the basin.

2. The framework of the Bucharest Declaration

Regional co-operation and co-ordinated action amongst many partners was and is needed to solve environmental challenges in the Danube River Basin. By the 1980s, Danube countries had become aware of the need to meet together to protect the environment of the Danube.

2.1. MAIN OBJECTIVES OF THE DECLARATION

Bucharest Declaration objectives fall into three primary categories: general, specific, and potential.

2.1.1. *General objectives include :*

- Water resources conservation and their rational use will be achieved by incorporating the prevention and control of Danube River pollution as an integral part of the national policy of Danubian countries, using approaches that will
  - sustain the supply of Danube River water,
  - take preventive measures against pollution, especially those created by hazardous substances and radioactivity;
  - take gradual steps toward pollution reduction keeping in mind the overall ecological needs for the Danube River, and
  - use self-monitoring by each riparian country with respect to their discharges.
- International long term co-operation based on multilateral and bilateral agreements to provide
  - systematic observation of the Danube River water quality;
  - the elaboration of common programs and methodologies for monitoring activities;
  - development of comparable data.

Thus, even by 1985 Danube water quality protection was linked with general river basin management. Some more recent concepts were also under consideration as indicated above and included overall ecological protection, the establishment of an international monitoring programme, transnational monitoring network, and international data comparison.

2.1.2. *Specific objectives*
Specific objectives are served by three main activities: monitoring, laboratory analysis and the possibilities offered for information management.

Monitoring objectives Water quality monitoring stations were to be located at transboundary sites, at the beginning and the end of the common reach, based on bilateral agreement.

Other monitoring sites could be established, if considered necessary, based on bilateral agreement. These might include sites upstream and downstream of main tributary discharges, downstream of large cities and reservoirs and at the main Danube River discharge into the Black Sea.

Laboratory objectives Laboratory objectives include the reciprocal exchange of information concerning sampling and analytical methods. The possibilities of using common analytical methods and automatic monitoring stations and system implementation were reviewed. Analysis of the analytical results of Danube monitoring programme was to begin not later than 2 years from 1985. There was to be a reciprocal exchange of information concerning the methodology used in observation correlated with reporting of accidental pollution and determination of measures for prevention.

Information management After 2 years of data processing the characteristics of the Danube River over the considered period of time will be determined and information will be exchanged concerning technical solutions for waste water treatment processing, analytical methods and water quality standards. Also for the specific objectives there are provisions with actual concepts: network design criteria for the emission/immission control (risk assessment and management), reliability of the laboratory data, and comparison of the data.

2.1.3. *Potential objectives*
Possible future objectives include the extension of the Declaration in a step-by-step process so that it becomes an international convention. Goals would provide for the protection, conservation and improvement of environmental conditions based especially on the increasing need for water pollution abatement.

2.2. BUCHAREST DECLARATION MONITORING PROGRAMME

During 1985-1987 a common programme was established and agreed to for the transnational Danube River monitoring network.

2.2.1. *Framework requirements*
The overall goal of the programme was to format and adopt the basic rules and requirements for the water quality monitoring activity within the frame work of the Bucharest Declaration in order to obtain comparable data, which could be used for Danube River water quality trend evaluation relevant to site and to time periods.

The participating countries included Austria, Bulgaria, the former Czechoslovakia, Germany, the former Yugoslavia, Romania, Hungary and the former Soviet Union.

2.2.2. *Monitoring stations, frequency, parameters*
Monitoring stations: There are 13 stations described in Table 1. Three relevant items should be mentioned:

- samples are collected by both countries based on an agreed time table;
- as a general rule the samples are taken from the left, middle and right side of the River;
- rate of flow is measured at the same time.

TABLE 1. Danube River quality monitoring stations

Nr.	Control Station	Country	rkm	Position
1	Jochenstein	Germany/Austria	2203.8	
2	Wolfsthal–Bratislava	Austria/Slovak Republic	1873.0	
3	Medvedev–Medve	Slovak Republic/Hungary	1806.4	
4	Cob–Szob	Slovak Republic/Hungary	1707.3	
5	Mohács–Bezdam	Hungary/Yugoslavia	1455.0	
6	Banatska–Palanka	Yugoslavia/Romania	1076.0	
7	Gruia–Radujevac	Romania/Yugoslavia	851.0	
8	Pristol–Novo-Selo	Romania/Bulgaria	833.6	Upstream of Jiu Upstream of Timok
9	Chiciu–Silistra	Bulgaria/Romania	375.0	Upstream of Jiu Downstream of Timok
10	Reni	Ukraine/Romania	166.0	Downstream of Prut
11	Vilcov	Romania/Ukraine	18.0	Chilia arm

Frequency: Sampling was accomplished at each site twelve times a year.
Parameters are identified in Table 2. There were three time periods anticipated:

I. an initial period from 1988 to 1990,
II. begun in 1991 to 1992 and
III. begun after 1993.

All the measurements refer to the total concentration (water column and suspended matter) without previous filtration for the particular determinands (in suspended matter or only in soluble forms).

The programme was organised generally into two time periods: (I) 1988-1990 as an experimental period and (II) after 1990 with improvement of methodology and the extension of the determinands range. Analytical methods follow those of the American Standard Testing Material of the American Water Works Association.

TABLE 2. The parameters monitored by the Bucharest Declaration activity

Cla	Definition	Determinands	Period of determination
1	General	(1) water temperature, (2) water transparency, (3) suspended matter	I
2	Dissolved oxygen regime	(4) D.O., (5) COD-Mn, (6) COD-Cr, (7) BODS	I
3	Salts - mineralisation	(8) conductivity, (9) T.O.S., (10) water hardness, (11) Cl, (12) S042, (13) Ca2+, (14) Mg2+, (15) Na+, (16) K+	I; for 12-16: II
4	Buffer capacity	(17) pH, (18) Alkalinity, (19) HCO3	I
5	Nutrients	(20) N-NH4+, (21) N-NOZ, (22) N-N03, (23) P-P043, (24) PT,(25) Chlorophill "a"	I, for 25: III
6	Heavy metals	(26) Fe, (27) Mn, (28) Zn, (29) Cu, (30) Cr6+, (31) Pb, (32) Cd, (33) H, (34) Ni, (35) As	II
7	Organic micropollutants	(36) Pherols (37) Detergents (38) Oil (39) AOX (40) Pesticides (O-Cl)	I I I III II-III
8	Microbiological	(41) Total coliforms (42) Total faecal coliforms (43) Total bacteria	II II II
9	Radioactivity	(44) a - Total alpha (45) (3- Total beta (46) Gamma spectrometry (47) Tr (48) Sr	III I II-III III III

2.2.3. *Organisation, data transmission*

The Environmental Research and Engineering Institute (ICIM) located in Bucharest, Romania has had the role of technical secretariat for Bucharest Declaration activities. As a general rule each riparian country transmits its collection of the previous year's data in April. Other important steps included:

- agreement that results be based on mean values;
- provision by ICIM of additional information concerning the Danube River water quality status and trends;
- provision of a special file for exchange of information in the case of accidental pollution (Doc. 3) with the following data:
  1. Date, hour and place of the accident
  2. Type and accidental pollution magnitude
  3. Measures taken
  4. A forecast of the pollution plume direction data and magnitude.

2.3. RESULTS OF BUCHAREST DECLARATION ACTIVITIES

International co-operation on water resources increasingly focuses on the prevention, control and reduction of transboundary water pollution, as well as on the ecologically sound and rational management of transboundary waters. The Convention on the Protection and Use of Transboundary Water Courses and International Lakes (Helsinki,1992) which was elaborated under the auspices of the United Nations Economic Commission for Europe (UNECE), has given such efforts, including those under the Bucharest Declaration, new impetus. The Convention covers surface water and ground water which mark, cross or are located on boundaries between two or more states, and their related ecosystems. At the regional level, it is the first legally binding instrument to prescribe measures for the prevention, control and reduction of significant adverse effects on the environment resulting from a change in the conditions of transboundary waters caused by human activity. These include effects on human health and safety, flora, fauna, soil, air, water, climate, landscape, and historical monuments or other physical structures among these factors.

Comprehensive knowledge about the state of water resources and related ecosystems is required to implement this convention. Moreover, monitoring compliance with its provisions demands reliable data on waters and factors influencing water quality and quantity. There is, for instance, a need for data related to in-stream quality, such as conditions of water quantity and quality, biota and sediment. Emission sources also have to be monitored to obtain information on the concentration of pollutants in effluents, and to carry out pollution load compilations. These general requirements are only partially covered by the Bucharest Declaration framework.

2.3.1. *Key requirements*

Five major interrelated requirements for monitoring and assessing the conditions of transboundary waters can be singled out: (i) appropriate design of monitoring programmes on waters; (ii) monitoring of significant emission sources; (iii) warning and alarm systems; (iv) harmonisation of approaches of monitoring and assessment at the regional level and (v) easy access to data. To this must be added the diversity of hydrogeomorphological units (HGMU) specific for the Danube River Basin.

A comparative point of view of the compliance with these key requirements based on the actual facilities given by Bucharest Declaration / Environmental Programme for the Danube River Basin and the Danube River Protection Convention is given in the Table 3.

TABLE 3. Danube River Basin Monitoring Key Requirements

Key requirements	Degree of compliance		
	Bucharest Declaration	EPDRB*	DRPC**
1⁰ WATER AMBIENT MONITORING (MISSION)	+	+	+
1.1. Investigated media :			
1.1.1. water column	+	+	+
1.1.2. suspended matter	-	- +	-
1.1.3. sediments	-	+	-
1.1.4. biocenosses	+	+	+
1.2. Chemical monitoring	+	+	+
1.2.1. concentration trend	+	+	+
1.2.2. load trend	+	+	+
1.2.3. compliance with WQC / WQO***	-	-	-
1.2.4. point pollution	+	+	+
1.2.5. diffuse pollution	-	-	-
1.3. Biomonitoring	+	+	+
1.3.1. saprobic index	+	+-	+ -
1.3.2. plankton	+	+	+
1.3.3. benthos (zoo)	-	+	+
1.3.4. biodiversity	-	-	-
1.3.5. ecosystem factors	-	-	-
2⁰.EFFLUENT MONITORING (EMISSIONS)	- +	+	- +
2.1. concentration	+	+	+
2.2. load	-	- +	- +
2.3. bioassay	-	+-	-
3⁰. NATIONAL POLLUTANTS BALANCE	-	+ -	+ -
4⁰. EARLY WARNING	-	+ -	+ -
5⁰. A SYSTEM INTEGRATED APPROACH Environmental functions / environmental themes / priority pollutants target groups	-	+ +	+ -

* Environmental Programme for the Danube River Basin

** Danube River Protection Convention

*** Water Quality Criteria / Water Quality Objectives

2.3.2. *Data access*

It has been agreed that the ICIM - Bucharest Institute provides the Bucharest Declaration Annual Report based on the data received from the riparian countries which contains a report of Danube water quality in the previous year (according to appendix 3 of the Declaration agreement on methodology). For each section the report presents :

1. DO, TDS, nutrients, specific pollutants, saprobity, microbiological parameters, radioactivity time trends (increase, decrease, stationary values) in the last 1- 2 years;
2. the overall quality of the sites compared with the previous year (generally stationary situation);
3. chemical mean values spatial trend;
4. mean loads values spatial trend;
5. annual statistical values (min., max.,mean, dev ±) for chemical determinands at each station.
6. monthly values for the chemical determinands at each station.

All reports are provided to each delegation. At the meeting it was proposed to use PC diskettes for the data exchange, based on a unique Data Exchange Format File (Bulgarian proposal).

The Bucharest Declaration data have also been given to the INFODANUBE database.

2.3.3. *Operational / operative aspects and needs*

Urgent needs include the following:

- There is need for an appropriate network design, the selection of other suitable monitoring parameters and the proper determination of sampling frequencies in order to prove that significance effects or human health and safety, flora, fauna, soil, air, water, climate, landscape and historical monuments or other physical structures do not occur. Existing programmes may not need all these requirements (integrated approach). The experience gained within the framework of programmes carried out in Europe (River Rhine, River Elba, etc.) could help countries with economies in transition (CIT) to cope with existing monitoring difficulties. Most of these countries suffer the consequences of past development. There is need for capacity building and for the transfer of technology on monitoring hazardous substances in water (including laboratory services), biological monitoring of water courses, monitoring of sediment quality and of radioactive substances.
- There is an urgent need for monitoring significant emission sources in catchment areas of transboundary waters and for documenting the results of such monitoring. These activities include both actual measurement and data reporting as well as documenting the characteristics of the current treatment technologies. Codes of practice - harmonised on the international level - could provide guidance on how to fulfil these responsibilities.
- There is a need for implementation of the AEWS (Accident Emergency Warning System - EPDRB) not only for the Danube River but within CIT transboundary waters regarding the creation of such services. Warning of extreme hydrologic events is established practice in most of the CIT. Practical experience on the setting-up and operation of systems to warn on water quality emergencies seems to be limited, however, to a handful of these countries.

There is a need to harmonise the rules for setting up and operating monitoring programmes, measurement systems, devices, analytical techniques, data processing and evaluation procedures and methods for registering discharged pollutants. Regional co-operation is clearly required to attain comparability of data validity and compatible data systems. The need to harmonise methods for assessing the status of Danube River and tributaries - or data evaluation procedures, as singled out above - is a particular component of "compliance monitoring". Although a number of assessment and classification systems are used or tried out in transboundary waters, they may not be applicable throughout the region. Moreover, a number of CIT indicated a need to replace the obsolete water - quality classification developed in the 1970s.

- There is a need to draw up reporting procedures, such as agreed formats for the exchange of data on transboundary waters and emission sources, in order to facilitate the access to, and the widest exchange of, information, as early as possible, on issues covered by the provisions of the Convention.

Most of the above mentioned needs are covered by the objectives of the EPDRB.

2.3.4. *Trend problems*

Some evidence of trends in the quality of the Danube should be mentioned. Recent data show, for example, that even through water quality remained in the second class along most part of the Danube River, there is clear evidence, mainly in some sectors along the lower reach, of a shift of water quality toward the third class.

It should be noted that the monitoring systems in the Danubian countries were designed exclusively for water quality assessment, taking into consideration mostly the basic physical, chemical and saprobic parameters. Before monitoring began under the Bucharest Declaration there was a significant difference between spatial and temporal organisation of sampling programmes, sets of parameters and methodologies. In spite of significant improvements in monitoring systems after 1985, through the use of a common set of parameters and sampling methods, the resulting data bases available were not satisfactory for decision makers especially for the lower Danube River and focused on the water quality of the main channel.

Despite the development of a new monitoring programme under the EPDRB which includes a number of improvements in sampling and data sharing, and so forth, it will probably remain unbalanced with respect to the evaluation of biological resources (including biodiversity), the fluctuations and functions of the catchment basin. We have to consider also that there are many gaps in our knowledge concerning the functioning (including fluxes along the three spatial dimensions of Danube River System), productivity and carrying capacity of most of the ecosystem types which can be identified within the river system.

According to the results obtained through the working programme carried out within the Bucharest Declaration, it is possible to assert that at present the Danube water quality is still generally in class 2-4. Then studies of the last 25 years however, show the dynamic character is becoming less favourable. This evolution is clearly brought into relief by the values of the oxygen regime determinands, mineralisation, nourishment elements and by the saprobic bioindicators.

3. Conclusions

The Bucharest Declaration represented the first international agreement to set in place common monitoring procedures for the protection of Danube River water quality. It took from 1985 to 1988 for the preparatory phase but after that the programme ran well.

Monitoring activities have focused on: (I) spatio-temporal Danube River water quality, (II) transferred loads (III) warning system in the case of the accidental pollution. Monitoring compliance has been realised by reporting the result for the stations from the previous year.

The Bucharest Declaration developed the first comparison of analytical data on a matrix.

At present the Bucharest Declaration monitoring represents a possible basic activity for the Danube Interim Commission. On the other hand the Bucharest Declaration represents also a suitable frame work for the Environmental Programme for the Danube River Basin implementation of monitoring. When the Bucharest Declaration activity becomes part of the monitoring activity of the Danube River Protection Convention it will represent important basic support for implementation of its goals.

References

1. Declaratia privind Colaborarea Statelor Dunarene in probleme de gospodarire a apelor Dunaii si, in special, de protctie a apelor fluviului inpotriva poluarii adoptata la Bucuresti - *Hidrotehnica nr. 1.*, **Vol. 31**, (1-2) 1986.
2. Action for a Blue Danube -1995
3. Convention on Cooperation for the Protection and Sustainable use of the Danube River (*Danube River Protection Convention*) - March 1994. 5.
4. Verbocket H. : Trend Detection *Add. Estimation of Loads, the implication for Network Design Monitoring tailoring Symposium* - 1995 Netherlands (47-48).
5. Enderlein R. - Vadineanu A.: *Transboundary Rivers and Lakes* ibid (9-10)
6. *Basic Requirements for the Management and Assessment of Large International Water Systems : Danube River - Black Sea System* ibid (23-24).
7. Rojanski V. - Varduca A.: *How clean is the Danube ? An overall view. Habitat Danube - on European eco-system.* Proceedings of the International Colloquium 19-21 apr.1994 at U/m (72-99).
8. Rojanski V. - Varduca A.: Tendinte in evolutia calitatii Dunarii *Hidrotehnica*, **Vol. 39**, 9, 13-30 (1994).
9. Rojanski V. - Varduca A.: Present state and technica - scientifically co-operation perspectives in the framework of the Bucharest Declaration of 1985. *Mediul Inconjurator* **Vol. IV**, nr.2/1993 - (33-38).
10. Varduca A. - Marculescu V.: Points of view reporting main substantial elements necessary for revising STAS 4706/88. *Mediul Inconjurator*. **Vol. V**, nr.2/94 (40-50).

11. Varduca A. : Bucharest Declaration - Convention on Cooperation for Protection and Sustainable use of Danube River Environmental Programme for Danube River Basin - Brno 1995.

THE DANUBE RIVER PROTECTION CONVENTION:
Environmental data-sharing provisions

MR. WILHELM KITTINGER
President
Interim International Commission for the Protection of the Danube River
Secretariat: Vienna International Centre
Building J, A-1140 Wien

KEYWORDS / ABSTRACT : Danube River Basin / Danube River Protection Convention / International Commission for the Protection of the Danube River Basin / environmental data-sharing

The implementation of the Danube River Protection Convention has had strong support from Danube Basin countries. Important to its success were such previous agreements as the Bucharest Declaration (1985), the Sofia Environmental Conference of 1989 and the Helsinki Convention on the Protection and Use of Transboundary Watercourses and International Lakes (1992). An Interim Secretariat has been maintained in Vienna to support negotiation processes which concluded with a signing conference in June 1994 in Sofia, Bulgaria. A transfer of responsibilities from the Environmental Programme for the Danube River Basin Programme Co-ordinating Unit is under way. The countries have agreed to share data with each other and the public about the environmental conditions of transboundary waters; emissions and water quality monitoring, focusing on the results of water and effluent monitoring and assessment as well as on results of compliance checks with water quality objectives and permit conditions.

1. Present stage, interim implementation and co-operation

A discussion of the current interim phase and the future implementation phase of the "Convention on Co-operation for the Protection and Sustainable Use of the Danube River" (Danube River Protection Convention, or DRPC) would not be complete without recalling its history. Reviewing its history also brings to mind the spirit of co-operation and good neighbourliness which prevailed as well as the strong encouragement and policy support on which the implementation of the Convention will be based.

I. Lyons Murphy (ed.), Protecting Danube River Basin Resources, 43–47.

Several previous developments and initiatives should be mentioned. The first such evidence of Danube co-operation, the Bucharest Declaration of 1985, provided a valuable European framework for the development of and negotiations about the Danube River Protection Convention (DRPC). The Bucharest Declaration initiated co-operation with respect to the environment in the Basin and started an important learning process. A new impetus was given by the Sofia Environmental Conference of 1989. The Convention on the Protection and Use of Transboundary Watercourses and International Lakes was signed at Helsinki on 17 March 1992, constituting an all-European framework (Helsinki Convention 1992) for this important issue.

The Draft Danube River Protection Convention was finalised in the course of only three negotiation rounds which took place in Vienna, Bratislava and Bucharest. Throughout the whole negotiation period an Interim Secretariat was maintained in Vienna to support negotiation processes. Two preparatory meetings at Sofia and Prague paved the way for the Signing Conference which was held 29 June 1994 at Sofia. At the same time a Ministerial Declaration was adopted as a basis for the Convention's implementation pending its entry into force. The day after the Signing Conference the International Commission was already established at Sofia at an interim basis.

This Ministerial Declaration at the same time requests the IC to closely cooperate with the Danube River Basin Environmental Programme (EPDRB) and its "Strategic Action Plan (SAP)" the implementation of which is also based on a Ministerial Declaration as adopted 5 December 1994 at Bucharest. The SAP provides the responsibility for the water-related parts of the EPDRB to be transferred to the International Commission (ICPDR) once the DRPC's entry into force is realised.

This transfer will be both a considerable input for preparing the DRPC's implementation and a great challenge. Great things will be expected from the International Commission during the follow-up of EPDRB activities. In the Danubian context they can only be achieved through a step-by-step process, based on priority considerations.

The limitation of human and financial resources is a basic factor in the implementation of the IC which must therefore strive for the best possible selection of areas of concern as well as efficiency. Duplication must be avoided wherever it seems likely to occur. This applies to the present interim phase as well as to the implementation after the entry into force.

It has been one of the initial goals for developing the EPDRB and the DRPC to take on manifold activities in the Danube River Basin which did create the possibility of duplication. Accordingly in the interim phase close co-operation between the EPDRB and the DRPC has avoided parallel activity.

2. Legal basis for transboundary information exchange and data sharing

In the field of water quality management as well as globally in the field of river basin management the legal basis for transboundary information exchange and data sharing available to the Danubian countries is twofold. The Helsinki Convention of 1992 stipulates applicable framework provisions. The Danube River Protection Convention of 1994 (DPRC) includes some provisions addressed to or associated with transboundary information exchange and data sharing as to specific fields of co-operation.

2.1 HELSINKI CONVENTION 1992

The Convention is divided into two parts. The first is addressed to all Contracting Parties, while the second contains provisions relating to Riparian Parties. This part requests them to define their mutual relations, conduct and co-operation with special regard for the prevention, control and reduction of unfavourable transboundary impacts.

The Riparians have to make available, upon a specific request, reasonably available data, inter alia, on:

- environmental conditions of transboundary waters;
- experience gained as to best available techniques and respective results of research and development;
- emissions (pollution sources) and water quality monitoring;
- measures taken and planned to be taken to prevent, control and reduce transboundary impact including permits or regulations, and
- planned activities likely to have transboundary impact.

Apart from this exchange between Riparian Contracting Parties information is also to be made available to the public. This information is focused on the results of water and effluent monitoring and assessment as well as on results of compliance checks with water quality objectives and permit conditions.

2.2. DANUBE RIVER PROTECTION CONVENTION 1994:

This Convention was developed in the framework of the 1992 Helsinki Convention where a clear mandate for developing transboundary co-operation is stipulated. The scope of the DRPC, which applies to the entire catchment area of the Danube River, in practice includes all water-related activities and issues. Some core provisions are specified in detail, others are contained as part of a framework in a more general or implicit way. It is up to the International Commission to develop this framework further by setting priorities and taking up actual issues as required by joint co-operation in practice.

In addition, this framework will provide for the International Commission's systems for internal data sharing as well as for a data bank and public information access. Such networks will take into consideration all aspects of information exchange. These provisions in particular will be concerned with the following activities and issues:

- the enumeration of groundwater resources subject to a long-term protection;
- periodic inventories of the relevant point and non-point sources of pollution within the catchment area of Danube River to monitor the progress made in the implementation of the joint action programmes by establishing periodical progress reviews;
- development of concerted or joint programmes for monitoring the riverine conditions concerning both water quality and quantity, sediments and ecosystems;
- development of joint or harmonised methods for assessing waste water discharges including processing, evaluation and documentation of data;
- reports to the International Commission on basic issues necessary for the Commission to comply with its tasks;
- an exchange of information between Contracting Parties within the framework of the International Commission;
- protection of information supplied must be guaranteed; this also applies to information supplied to the public the conditions of which are specifically formulated;
- the results of joint or other relevant research, the access to which is open for public authorities, are to be transmitted to the Commission.

3. The Data Management Sub-Group (DMSG) and the Danube Information System (DANIS)

The DRPC's provisions addressed to or associated with information exchange are referred to in more detail in this presentation. This is in order to demonstrate that the future International Commission's activities will not entirely be subject to public information exchange, for technical and formal reasons.

The DANIS system originally was built up as a medium for exchanging institutional information. At the 9th Task Force Meeting of the EPDRB held at Sofia in March 1996, the Data Management Sub-Group proposed that this system be considerably extended to become a comprehensive data sharing system. It was proposed that sectoral data bases such as for monitoring, emission inventories and a project data bank become part of the DANIS system.

The IC expressed its concern based on the DRPC's provisions referred to above. For technical reasons raw data should not be introduced into the DANIS System before being evaluated and summarised. Moreover the IC's internal reporting activities must be reflected in summary reports adopted by the IC before becoming an input to the system. If in individual cases more detailed information would be required, DANIS could indicate where to get such information.

The most decisive point of hesitation expressed by the IC was its concern as to the financing problems the extended DANIS concept might imply. As this concept is oriented on a future transfer of responsibility to the IC and on the respective further development of the DMSG and the DANIS-system, the IC is highly interested in a realistic estimate of the follow-up costs for operating the system at the short-, medium- and long-term scale. If the transfer proves to be feasible, and no other more efficient less costly system found, the system will be taken up and extended step by step by the IC. If other approaches prove to be less costly and provide more information to a larger number of people they need to receive serious consideration.

These considerations must be understood in a very informal way without prejudicing the IC's deliberations on this subject.

Consequently also the President's views as to a possible NATO-data sharing system are expressed representing the IC's interests without prejudice to its future position.

Anyway the IC's situation regarding its human and financial resources as well as the difficult economic conditions and the domestic budgetary problems, which most of the Contracting Parties have to face, do require that we strive for the best possible efficiency and to avoid any duplication. This applies both to the implementation of the DRPC as well as to the responsibilities to be transferred from the EPDRB to the IC as far as water-related issues are concerned. When co-operating with other international organisations concerning water-related activities within the Danube River Basin the IC should seek to achieve the same goals.

WATER QUALITY MONITORING PROGRAMMES FOR RIVERS IN GERMANY

Prof. Dr. HANS LIEBSCHER
Federal Institute for Hydrology
Koblenz, Germany

KEY WORDS / ABSTRACT: water quality data-sharing / water quality monitoring / German water quality monitoring programme

A series of water quality monitoring systems have been established in Germany in recent years. The German Federal States are responsible for general water resources management and have created an extensive monitoring network for water quality control of ambient waters. Ground water monitoring is presently being installed. Since the Federation is responsible for traffic-related water management needs monitoring networks maintain records of river training measures. Some water quality stations are integrated with international river section commissions such as the International Commissions for the Protection of the Rhine and Elbe. Specific measuring programmes for coastal water quality routinely monitor the North and Baltic Seas using aeroplanes especially equipped with remote sensors.

1. Introduction

The advancing industrialisation in the course of this century was accompanied in Germany, as it was elsewhere, by massive emissions of anthropogenic contaminants into the atmosphere, to soil, and into waters. The main impairments of waters originate above all from discharges of domestic and industrial wastewaters and from diffuse inputs of substances from agricultural and urban areas. Negative effects result for the "protection targets", drinking water, aquatic communities, sediments, as well as uses of waters for recreation and sports. This gave rise to the necessity to monitor the ecological state of waters, the aquatic phase, their sediments and suspended solids, and selected organisms (biota). Pollution loads and trends had to be determined and assessed with a view to the effects of emission reduction and issues of human health, and point sources of pollution had to be identified. By providing a basis for political decision-making, these monitoring programmes have contributed to reductions of classical pollution loads such as heavy metals and chlorinated hydrocarbons in most German rivers and coastal waters.

Monitoring consists in regular observations at selected sites to collect, process, and evaluate data on water quality. A particular role herein is due to special monitoring

I. Lyons Murphy (ed.), Protecting Danube River Basin Resources, 49–64.

programmes of limited duration on topical issues as well as for preventive action and the collection of analytical evidence,

The protection of waters in Germany is the responsibility of the Federal States (*Länder*). For this reason the Federal States have *inter alia* monitoring programmes for water quality in watercourses. The Federal States have pooled together to form the Joint Water Commission of the Federal States (LAWA) which has *inter alia* the task to define common objectives in water quality monitoring. An essential precondition for the uniform evaluation of data and the presentation of the quality status of waters in all Federal States is a homogeneous data base. However, different emphasis on monitoring criteria and, in part, differences in staffing and technical equipment in various Federal States result in differing monitoring programmes.

To make possible the assessment of water quality by comparable criteria between the Federal States and to meet the growing requirements in the national and international exchange of data, it is necessary to collect water quality data in all Federal States at selected stations according to a standardised observation programme. For the harmonisation of the concepts of the Federal States, the LAWA working party "Qualitative Hydrology of Watercourses" has formulated the "Empfehlungen für regelmäßige Untersuchung der Beschaffenheit der Fließgewässer in den Ländern der Bundesrepublik Deutschland" (Recommendations for regular monitoring of the quality of watercourses in the Federal States of the Federal Republic of Germany).

As early as in the mid-1950s, Germany began to establish monitoring programmes for inland watercourses and marine waters. Since the mid-1960s, the Federal States have intensified the installation of new water quality monitoring stations. New knowledge about the pollution of waters and advances in measuring technology have contributed to the enormous expansion of water quality monitoring in recent times.

Altogether 150 LAWA stations have been established so far (1995) (Fig. 1.). This figure includes EC stations and stations operated by international river protection commissions and by several joint programmes of Federal States.

2. Objectives

The present water quality monitoring programmes serve the following purposes:

(1) Determination of long-term developments;

- Mean- and long-term determination of the quality of watercourses forms the basis for a nation-wide water quality inventory and for the identification of long-term tendencies. Suitable measuring strategies allow the recognition trends in water quality and to initiate counteractions if needed.

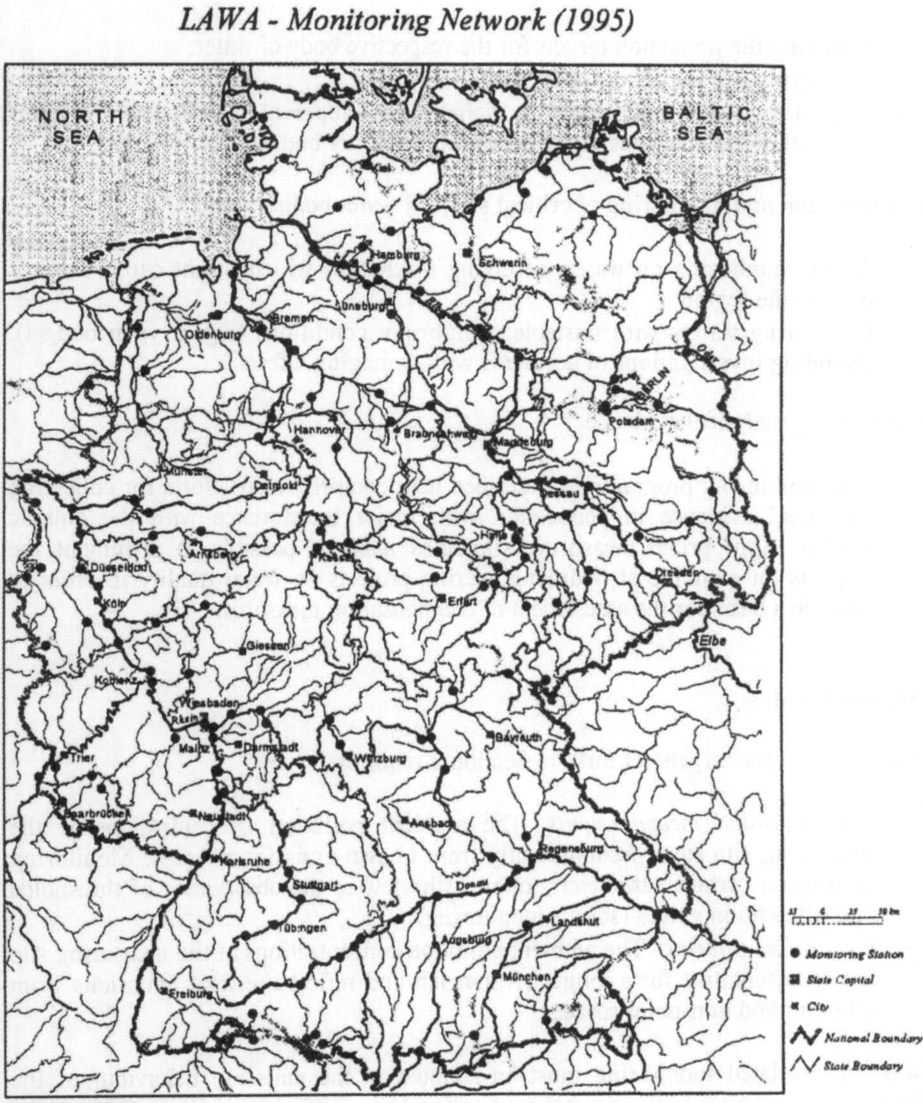

Figure 1. Water quality measuring stations in Germany

(2) Providing the basis for planning in water resources management

– Collections of water quality data are decision aids and the planning basis for water resources management, e.g. management plans, discharge permits and the like. Measuring strategies must permit the derivation of fundamental data for certain water bodies for planning purposes and administrative decision making.

(3) Checking the observance of water quality requirements, which includes:

- Checking the protection targets for the respective body of water;
- Identifying the impact of water uses;
- Monitoring restoration measures, including development of sanitation measures and ranking of priorities in the restoration of water bodies.

(4) Observation and monitoring of critical states of water bodies, including

- Alarm monitoring on waters receiving discharges which might cause damage outside the region;
- Monitoring waters with unstable equilibrium conditions (e.g. oxygen budget), including investigations of accidents with damaging effects.

(5) Collecting analytical evidence

- The monitoring programmes described here comprise also efforts for collecting analytical evidence in connection with major interference with thc aquatic system. Appropriate measuring strategies make it possible to document the impacts for example of hydroengineering projects on water quality in order to provide a basis for the assessment of compensation measures.

3. Measuring sites

Achievement of the targets set forth in Section 2 requires:

- *Object-related measurements*: The ambient pollution concentrations at the measuring site stem predominantly from known emission sources. Monitoring focuses on certain parameters, often with view to the observance of thresholds and orientation values (source monitoring).
- *General monitoring*: The ambient pollution concentrations at the measuring site are representative for a longer river reach and reflect the total emissions from all point and non-point sources.

While object-related monitoring must be adapted to the emission behaviour of the pollution source (water user), general monitoring has a different orientation. Here, the tasks are more or less concentrated, on two targets:

- Discharger-oriented water monitoring
- An overall estimate of the status of the water body.

The measuring sites are usually sampling points. If necessary, they are also equipped with measuring and sampling devices. It is essential that streamflow at the site is known, either from direct measurements or derived from data of neighbouring gauging stations.

The monitoring stations of the Federal States are sited in such a way that all major watercourses are covered, that impacts of state-wide relevance are identified, and that anthropogenic as well as the geogenic effects can be determined. This also includes the consideration of the importance of the watercourse for water resources management, the structure of the landscape and its natural endowment, the pollution load. and the size of the catchment area.

Monitoring stations of the Federal States are usually placed

- before the mouth of rivers of relevance for water resources management into the sea or coastal waters;
- on transboundary waters near borderlines;
- on important reaches of larger watercourses;
- on major tributaries immediately upstream of the inflow;
- on reaches without anthropogenic impacts ("zero stations", background stations).

The monitoring site should be as representative as possible for the river cross section (no inhomogenous distributions of temperature and concentrations).

4. Measurements

Water quality data are acquired by means of accepted analytical methods according to the standards DIN (Germany), ISO, or CEN in order to ensure comparability of results. The LAWA Commission formulated the "Rahmenempfehlungen der Länderarbeitsgemeinschaft Wasser (LAWA) für Wasser-, Abwasser- und Schlammuntersuchungen" (General recommendations of the Joint Water Commission of the Federal States for analyses of water, wastewater, and sludge) (1991), which include rules for analytical Quality Assurance (QA). LAWA issues regularly QA information sheets for certain parameters or measuring methods.

The following measurements are made (physical, chemical, and biological parameters):

- in situ;
- at measuring stations;
- in the laboratory.

5. Measuring programmes

Three categories of monitoring strategies can be distinguished. First, chemical-analytical methods which determine the concentrations of known contaminants;

second, eco-toxicological methods which are characterized by the use of bioassays and toxicity tests or investigations of biomarkers; and third, ecosystematic approaches which examine the development of the environment by means of important structures and functions of ecosystems including human activities. All these approaches have characteristic advantages as well as drawbacks, which are described in brief below:

Chemical-analytical methods provide quantitative information on individual substances and on summative parameters. The results can be compared with thresholds or quality targets for individual substances. Today more than 100,000 chemicals are in daily use and are emitted into the aquatic environment. Only a very small portion of them (<0.2%) is measured under monitoring programmes. Merely 5 - 10% of all known synthetic chemicals have been examined by toxicology so far. Chemical substances may undergo biotic and abiotic transformation processes and may form decomposition products (metabolites), which are often more toxic than the original substances but are not covered by the analyses applied.

Synergistic or antagonistic effects between contaminants and contaminant classes cannot be monitored. Chemical analyses do not give any information about the bio-availability of contaminants. Biotests make possible integrating statements about the effects of contaminants. They also cover such substances that are presently not analysed chemically. However, in the presence of complex mixtures of substances, as they are common in environmental samples, it is not possible to identify substance-effect relationships. The transfer of findings from laboratory tests to field conditions or from test organisms to local species is very problematic.

Ecosystem approaches permit observations to be made in the natural landscape units in their entity and to include socio-economic aspects. The distinction between natural and anthropogenic effects can be found by additional causal research. The assessment of environmental changes, predictions on the consequences of human interference, and well-founded future scenarios are only possible on the basis of steadily increasing knowledge about these systems.

These studies are very expensive in terms of organisational effort and finances. The interpretation of the findings needs particular expertise.

Present monitoring techniques are strongly oriented towards the determination of individual substances. The results of such investigations are "snap shots" which reflect the load of individual substances in the examined matrix. In Germany efforts are being made to bundle the parameters in various monitoring programmes. These cover investigations of water, suspended solids, sediments, and organisms, as well as aquatic communities and the aquatic, amphibious, and terrestrial habitats of water bodies Figure 2 shows a setup of a water quality monitoring station.

For water quality studies the relevant parameters are then concentrated in various separate sub-programmes. The parameters needed in order to evaluate the oxygen and nutrient budgets, for instance, are summarised in a basic chemical measuring programme (C). In 1993, LAWA published "Karten der Wasserbeschaffenheit - 1982-1991" (Maps of water quality - 1982-1991) which shows the trends of various parameters at LAWA monitoring stations. This survey gives additional information beyond the conventional water quality map which is based on biological-ecological criteria (cf. LAWA: Die Gewässergütekarte der Bundesrepublik Deutschland - The water-quality map of the Federal Republic of Germany). The basic biological

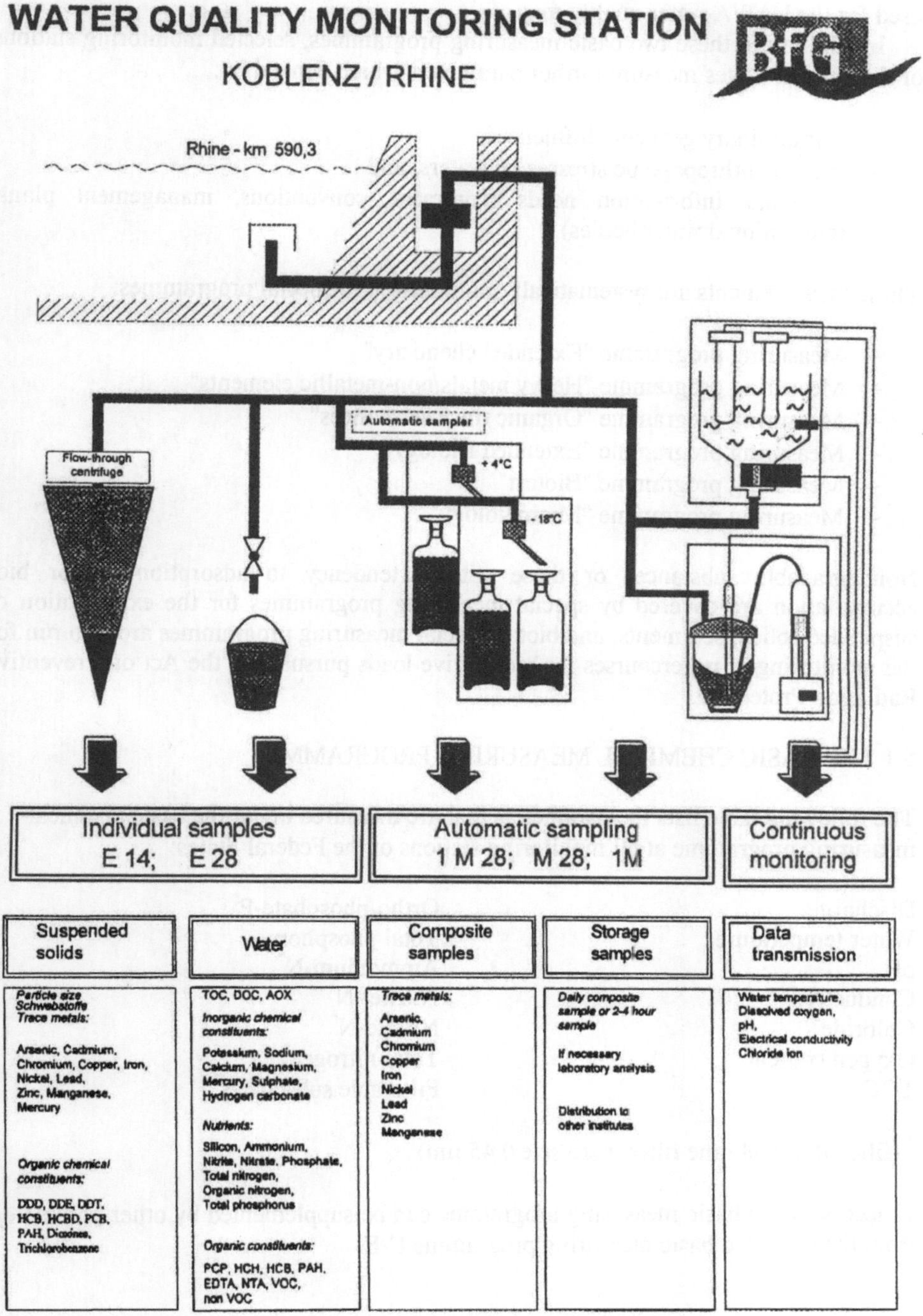

Figure 2. **Setup of a water quality monitoring station**

measuring programme (Bio) corresponds to the procedure for biological quality studies used for the LAWA water quality map.

In addition to these two basic measuring programmes, selected monitoring stations of the Federal States measure further parameters which depend on:

- extraordinary geogenic influences;
- special anthropogenic stresses on waters, and
- particular information needs (contracts, conventions, management plans, international water bodies).

These measurements are systematically summarised in special programmes:

- Measuring programme "Extended chemistry"
- Measuring programme "Heavy metals/non-metallic elements"
- Measuring programme "Organic trace substances"
- Measuring programme "Extended Biology"
- Measuring programme "Biotest"
- Measuring programme "Microbiology".

Nondegradable substances or those with a tendency to adsorption and/or bio-accumulation are covered by special measuring programmes for the examination of suspended solids, sediments, and biota. Special measuring programmes are also run for the monitoring of watercourses for radioactive loads pursuant to the Act on Preventive Radiation Protection.

5.1 THE BASIC CHEMICAL MEASURING PROGRAMME

The following table lists the parameters that are measured under the basic chemical measuring programme at all monitoring stations of the Federal States:

Discharge	Ortho-phosphate-P*
Water temperature	Total phosphorus
pH	Ammonium-N*
Conductivity	Nitrite -N*
Chloride	Nitrate-N*
Oxygen content	Total nitrogen
TOC	Filterable substances*

* (filtered, membrane filter, pore size 0.45 μm)

If necessary the basic measuring programme can be supplemented by other parameters from the extended basic measuring programme C-E.

5.2 EXTENDED BASIC CHEMICAL MEASURING PROGRAMME (C-E)

Parameters which are analysed under extended measuring programmes for special purposes (e.g. fisheries) include

DOC	Sodium
AOX	Potassium
BOD/oxygen consumption	Calcium
UV absorption at 254 nm	Magnesium
UV absorption at 436 nm	Aluminium
Sulphate	Dissolved total Nitrogen*
Hydrogen carbonate	Dissolved total Phosphorus*
Silicate*	(COD)

* (filtered, membrane filter, pore size 0.45 μm)

5.3 MEASURING PROGRAMME FOR HEAVY METALS/NONMETALLIC ELEMENTS

The following heavy metals are analysed as total contents, load on suspended solids, and dissolved portions after separation (filtration, centrifugation):

Lead	Mercury
Cadmium	Zinc
Chromium	Iron
Copper	Manganese
Nickel	Arsenic

For heavy metals which are attached in high percentages to suspended solids, the analysis can be made in suspended particulate material. Moreover, the analysis of biota (like algae, fish, mussels) may indicate the enrichment of heavy metals in aquatic organisms.

5.4 MEASURING PROGRAMME FOR ORGANIC MICROPOLLUTANTS

When organic micropollutants can be expected to occur (e.g. with increased AOX levels, special effluent discharges), a differentiating analysis of the individual substances helps to establish the relation to the emission source and to assess the risk with view to the respective protection target (toxic relevance). As a rule, at selected measuring stations relevant organic pollutants are investigated in special measuring programmes in their respective catchment areas.

Presently, selected river monitoring stations in the Federal Republic of Germany measure 28 substances for which protection targets have been defined, as well as EDTA and NTA. Analyses of organic substances of relevance for whole river basins are also performed in the context of measuring programmes of national or international river basin commissions.

Further differentiating analyses of suspended solids, sediments, and biota can be made to determine the enrichment of organic contaminants in the ecosystem.

5.5 BASIC BIOLOGICAL MEASURING PROGRAMME

Studies for the determination of the saprobic index for the classification of water quality (LAWA water quality map) are made according to the German standard DIN 38410.

5.6 EXTENDED BIOLOGICAL MEASURING PROGRAMME

Depending on requirement, supplementary biological examinations can be performed, such as

Chlorophyll-a;
Phaeophytin;
Qualitative and quantitative plankton studies;
Plankton biomass.

5.7 MEASURING PROGRAMME BIOTEST (TOX)

Depending on the objectives of the examinations, biotests with various test organisms (like fish, entomostracans, bivalves, algae, bacteria) are used, either as "static" laboratory tests or continuously as "dynamic" alarm tests in monitoring stations.

Tests for the determination of suborganismic effects of water constituents (e.g. cell tests, enzymatic tests) and tests for genotoxic effects have remained limited to special studies so far.

5.8 MEASURING PROGRAMME MICROBIOLOGY

For special uses of waters (e.g. for bathing, irrigation, cattle watering) or for the control of epidemics, bacteriological analyses are performed regularly or seasonally, covering for instance

Colony counts;
Total coliform counts;
Faecal coliform counts;
Faecal streptococcus counts;
Salmonella;
Viruses.

5.9 RADIOACTIVITY

The monitoring of watercourses for radioactive substances plays a special role. Here, the Federal authorities have the responsibility for the large-scale monitoring of the Federal waterways for radioactivity from normal operation and from incidents.

Radioactivity monitoring in the other surface waters is a responsibility of the Federal States.

Networks of sampling stations are operated to take samples of water, suspended solids, and sediment from watercourses and to analyse them for radioactive material. The instruments measure continuously the total-beta and total-gamma activities in ambient water for early warning purposes including staff alert.

Under normal conditions these studies can show especially the long-term effects of emitters and sources of radioactive material. In the case of nuclear incidents it is essential to recognise the increased discharge of radioactive substances immediately and to identify the hot-spots of radioactive deposition, so that the possible impacts can be assessed and recommendations for preventive actions to protect the population can be issued as soon as possible. The monitoring network of radioactivity on Federal waterways consists of 40 warning stations (Fig. 3.). In such monitoring stations the water is continuously pumped through a shielded measuring basin where a NaI (Tl) scintillation detector and a flat proportional counter record simultaneously the total-beta and total-gamma activities (Fig. 4.). If the pre-defined warning or alarm thresholds are exceeded, or if malfunctions occur, the stations automatically send a message to the central network computer where the information is evaluated and recorded. When the alarm threshold of the total-gamma measuring canal is exceeded and the computer has performed a plausibility check, the operating staff is automatically alerted.

Within the network samples of water, suspended solids, and sediment are taken. Sampling sites for water and suspended solids are often combined with warning stations so that the existing facilities can be utilised.

6. Sampling strategy

Two types of sampling for completion of the task are common, composite samples and individual samples.

6.1 COMPOSITE SAMPLES

Composite samples are usually taken proportionally in time, although today discharge-proportional composite sampling is technically feasible. In composite sampling, single samples are taken over a certain time to represent the whole sampling period or the cumulative discharge during this time. This type of sampling is preferred when the task is to determine annual transport rates exactly. The monitoring stations are equipped with automatic samplers. Sometimes, alterations in the contents of degradable contaminants during sampling pose problems, so that special sample preservation techniques are needed.

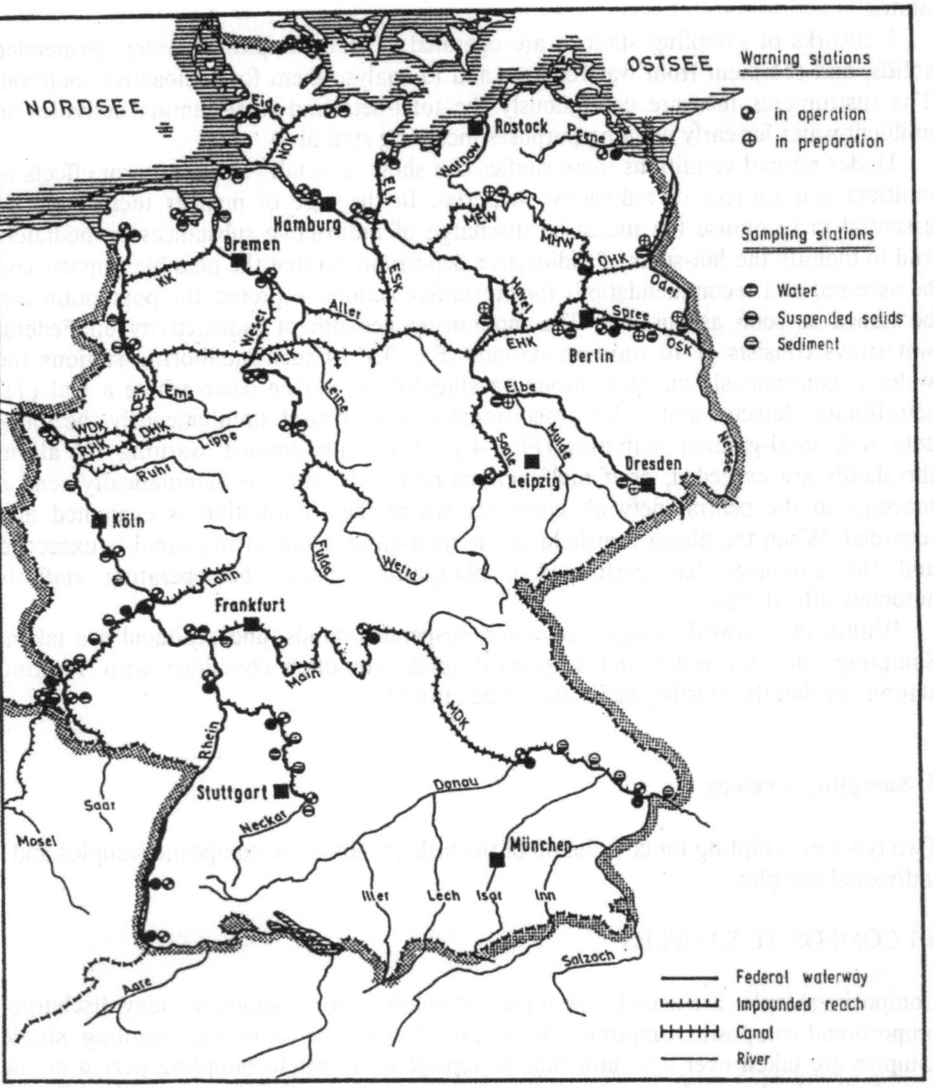

Figure 3. **Warning and sampling stations for measuring radioactivity at federal water ways**

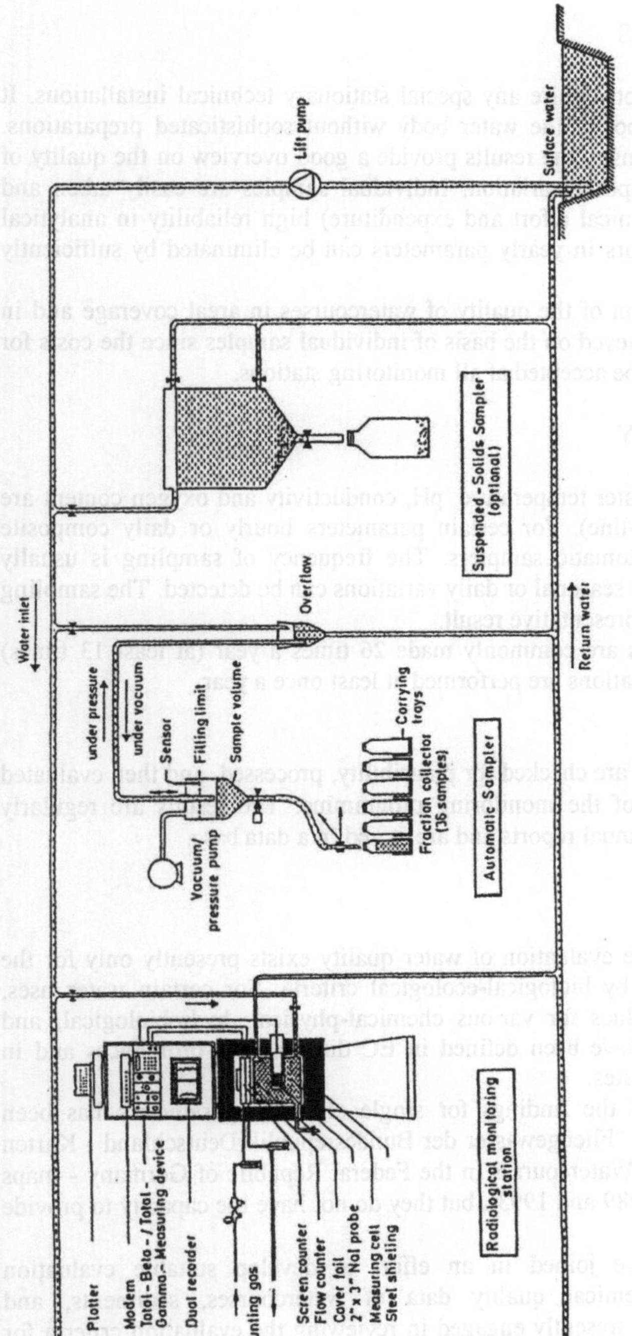

Figure 4. **Setup of radiological monitoring station**

6.2 INDIVIDUAL SAMPLES

Individual sampling does not require any special stationary technical installations. It can be performed at any spot of the water body without sophisticated preparations. With sufficient sampling density the results provide a good overview on the quality of the water body and its temporal variation. Individual samples are easily taken and provide (without major technical effort and expenditure) high reliability in analytical terms. Larger statistical errors in yearly parameters can be eliminated by sufficiently large numbers of samples.

A comparative description of the quality of watercourses in areal coverage and in its temporal variation is achieved on the basis of individual samples since the costs for composite sampling cannot be accepted at all monitoring stations.

6.3 SAMPLING FREQUECY

Some parameters such as water temperature, pH, conductivity and oxygen content are measured continuously (on-line). For certain parameters hourly or daily composite samples are taken with automatic samplers. The frequency of sampling is usually chosen in such intervals that seasonal or daily variations can be detected. The sampling sequence should ensure a representative result.

Chemical water analyses are commonly made 26 times a year (at least 13 times) and sapro-biological examinations are performed at least once a year.

6.4 DATA PREPARATION

The recorded measured data are checked for plausibility, processed, and then evaluated according to the objective of the monitoring programme. The results are regularly published in yearbooks or annual reports and are stored in a data base.

6.5 DATA EVALUATION

A binding regulation for the evaluation of water quality exists presently only for the water quality classification by biological-ecological criteria. For certain water uses, threshold or orientation values for various chemical-physical, hydrobiological, and bacteriological parameters have been defined in EC directives or ordinances and in regulations of the Federal States.

A first interpretation of the findings for single chemical parameters has been provided in the publications "Fließgewässer der Bundesrepublik Deutschland - Karten der Wasserbeschaffenheit" (Watercourses in the Federal Republic of Germany - maps of water quality) (LAWA, 1989 and 1993), but they do not have the capacity to provide detailed information.

The Federal States have joined in an effort to develop suitable evaluation procedures for physico-chemical quality data of watercourses, sediments, and suspended solids. LAWA is presently engaged in reviewing the evaluation criteria for water quality in the Federal states and compiles on this basis a recommendation for the evaluation of water quality by chemical-physical criteria.

7. Future developments

Adequate and cost-efficient monitoring and evaluation techniques exist only in some cases. The success of these approaches remains to be seen.

Monitoring programmes which include chemical-analytical methods have produced a confusingly vast amount of data which gives little information about the actual state of the environment.

Only by integrating the above-mentioned approaches with simulation models and by combining them with geographic information systems (GIS) and with remote sensing can these disadvantages be avoided and the costs of monitoring reduced. Such an approach has already been considered in some of the new ecosystem approaches.

The objectives of future monitoring strategies include:

- A multidisciplinary approach must comprise physical, chemical, and biological parameters.
- The measured data must be combined with socio-economic parameters in a management system.
- It must be possible to derive the relationship between environmental impacts and human interference (e.g. effluent discharges, fishery). Scenarios should identify options for political actions.
- The methods must be cost-efficient.

In this context the question arises as to how promising the monitoring concepts which show promise should be arranged in the future, which components are already available today, and which still must be developed. The chosen classification of substances which is in use needs to be reviewed to see if it is suitable for complete chemical-physical monitoring with environmental relevance.

Another decisive question requiring an answer is how chemical-analytical data can be acquired at low costs, and how their information value can be increased, e.g. by integrative analyses.

Moreover, we need to clarify which possibilities are opened by bio-markers as monitoring parameters. In this context the question whether chemical-analytical methods and toxicity tests can be combined to derive information on the effects of individual compounds is gaining in importance. It should also be examined how monitoring programmes, ecosystem research, and causal research can be combined in meaningful ways. A special role can be played by simulation models. Here, consideration should be given to the use of models to select the "proper" monitoring parameters.

Statistical methods are essential for data evaluation. A precondition for the application of these methods are reliable consistent data series. We must discover how historic and present-day observations can be combined into long-term data series. In particular, the suitability of the use of geostatistical methods for evaluations should be studied.

References

1. Deutscher Verband für Wasserwirtschaft und Kulturbau (DVWK) (1982): Meßstationen zur Erfassung der Wasserbeschaffenheit von Fließgewässern - Einsatz, *Bau und Betrieb - Merkblätter zur Wasserwirtschaft* 201, Hamburg (Parey).
2. LAWA (1991): Rahmenempfehlung der Länderarbeitsgemeinschaft Wasser (LAWA) für Wasser-, Abwasser und Schlammuntersuchungen.
3. Länderarbeitsgemeinschaft Wasser (LAWA) (1993): Karten der Wasserbeschaffenheit. 1982 - 1991.
4. Länderarbeitsgemeinschaft Wasser (LAWA) (1996): Fließgewässer der Bundesrepublik Deutschland - LAWA-Untersuchungsprogramm in den Ländern der Bundesrepublik Deutschland (in Vorbereitung).
5. Umweltbundesamt (UBA) (1994): Daten zur Umwelt 1992/1993.

DUTCH (INTER)NATIONAL WATER QUALITY DATA PROGRAMMES

DR. H. A. G. NIEDERLÄNDER
International Centre of Water Studies
Damrak 83
1012 LN Amsterdam, the Netherlands

KEYWORDS/ABSTRACT: Netherlands / water quality monitoring / water quality data / information sharing / international agreements

Netherlands water quality monitoring and data-sharing policies must respond to national as well as international commitments. Legislation is cited and summarised regarding water quality management and control as they reflect two complementary policy lines: the emission and the water quality based. Data and information on water quality are required for the preparation and evaluation of national policy objectives and strategies as laid down in legislation. On a national level, the results of water quality assessment and the assessment of discharges to surface water are periodically published in a number of reports. The more important report series are described, emphasising the importance of data-sharing from the national as well as the international point of view.

1. Introduction

The Netherlands is situated in the delta of three transboundary rivers: the Rhine, the Meuse, and the Scheldt. The Netherlands is also part of the North Sea catchment area. This situation serves to explain that the need for collecting water quality data does not only stem from juridical obligations and requirements for the preparation and evaluation of national policy objectives and approaches. Need is also derived from international programs like the Rhine and the North Sea Action Program (RAP/NAP). Furthermore, since the Netherlands is a member state of the European Union (EU) the requirements laid down in EU policy, including reporting obligations, are incorporated into national legislation and/or monitoring programs. These various information requirements are outlined in more detail below.

Various reviews of Dutch environmental legislation and policy approaches, including the administrative and managerial structure installed for implementation and control of associated requirements, have been consulted for the preparation of this paper. Some of these sources have not been explicitly quoted in the text; these include: [VROM, 1991], [OECD, 1995], [VROM, 1989], [RIZA, 1993].

I. Lyons Murphy (ed.), Protecting Danube River Basin Resources, 65–80.

2. Information requirements

In the Netherlands, water quality management and control are regulated along two complementary policy lines: the emission based and the water quality based. The framework for the emission based approach is laid down for the most part in the "Pollution of Surface Waters Act " (WVO) adopted in 1969 and amended several times [State Journal, 1992, 628].

This act prohibits any emission to surface water, and in some cases to sewer systems, unless a license has been issued. In accordance with EC legislation, the emission of "black list" substances (EC list I substances: directive 76/464/EEC and succeeding directives for specific substances) is prohibited and licensee will require Best Available Technology (BAT) to be issued in case of emission of these substances. Criteria for these substances included in the licenses have been based on BAT and should be at least as stringent as issued by the EC (i.e. for priority substances covered by a subsequent directive of directive 76/464/EEC). More stringent emission limits can be imposed if the protection of specific functions of surface water requires (water quality based approach).

This provision has been amended under the "Pollution of Surface Waters Act" by means of a General Administrative Order (AMvB) [State Journal, 1983,606], implementing EC legislation for bathing waters, surface waters used for drinking water, fishing and shellfish waters.

Criteria for non-black list substances, also aiming for the protection of these specific functions of surface water or based on guidelines (e.g. on toxicity, biodegradability, etc.) included in the "Environmentally Hazardous Substances Act" [State Journal, 1992, 632], can be included in the licenses as well. Licenses will require Best Practicable Means (BPM) to be used in case of emission of such substances. Moreover, the "standstill" principle is applied, meaning that in each administrative area the overall discharge of any particular substance must not increase. Similar arrangements for the protection of ground water are covered by the "Soil Protections Act" [State Journal, 1986, 374].[1]

Apart from criteria, requirements for self-monitoring of discharges by the emitting company are included in the license. In addition monitoring of the discharges is performed by the responsible authorities; both on a regular and ad hoc basis (without prior notification). Data from these monitoring efforts are used for checking of compliance with license agreements, and for charging purposes.

The "AMvB" for the protection of specific functions of surface water requires in addition monitoring of surface waters assigned such a specific function by the responsible authorities. Variables, criteria and minimum monitoring frequencies and analytical methods are an integral part of the "AMvB". However this framework allows only for water quality evaluation, on an annual basis, after the event. Direct operational protection is taken care of by such legislation as the "Hygiene and |Safety of Swimming Waters Act" [State Journal, 1982, 494] and the "Mains Water Supply Act" [State Journal, 1984, 220]. Data collection in the context of these and other "early warning" provisions that require water quality monitoring, however, has not been included in this paper.

3. National Policy Objectives

Apart from the juridical requirements for data collection described above, information on water quality is required for the preparation and evaluation of national policy objectives and strategies. The basis for the current national policy approach was laid down in the "Third Water Management Document" [V&W, 1989]. The system of intermediate and ultimate-quality targets for both ground and surface water, sediment and soil, as described in this document, was made operational, with minor changes, by notification through the Minister of Housing, Physical Planning and Environment in the document: "Environmental Quality Targets for Soil and Water" (MILBOWA) [VROM, 1991(a)]. Additional modifications were recently made in the "Evaluation Note Water Management" [V&W, 1994][2].

The system of intermediate and ultimate quality targets, installed through these policy documents, does not serve any direct juridical purpose. The aim is merely to set the policy goals for the short and medium long term and to provide a valuable tool to test the efficiency of current policy approaches. Nevertheless responsible parties have an obligation to make an effort toward reaching these targets within a predetermined period of time through the documents cited above. This obligation includes the requirement to assess water quality with respect to target compliance. Furthermore, it should be noted that these policy targets are invariably at least as stringent as the criteria for ambient water quality imposed by means of juridical provisions for special functions.

3.1 WATER QUALITY MONITORING PROGRAMMES

Routine water quality monitoring programmes in the Netherlands are highly structured towards an assessment based on the system of intermediate and ultimate-target values. Moreover, many other data collection and information reporting obligations can be fulfilled through these same monitoring programs. As such, knowledge of this system is of major importance with respect to administrative structures, monitoring and assessment activities and data sharing. A more detailed discussion of the concepts applied to define intermediate and ultimate water quality targets follows.

3.1.1. *International Agreements*

Further information requirements follow from agreements made within the context of various international programs for the protection of specific water bodies against pollution (Rhine and North Sea[3]) and member state obligations to report specific water quality data to international organisations (e.g. the European Commission).

Within the context of the Rhine Action Programme (RAP) [ICPR, 1987 and Rhine Ministers Conference, 1989], and the North Sea Action Programme (NAP) [North Sea Ministers Conference, 1990], agreements have been made concerning the reduction of emission loads of priority substances to these water bodies with a predetermined period of time (1985-1995). Assessing the effectiveness of policy approaches, installed to meet these agreements, requires data collection through inventory and monitoring. In addition, ambient water quality assessment programs (annual basis) are being agreed upon within monitoring working parties, operating under the International Commission for Protection of the Rhine against Pollution (ICPR)[ICPR, 1994] and the Oslo and Paris Commissions (OSPARCOM) [OSPARCOM, 1990]. Within the context of both programmes, background concentrations of naturally occurring compounds have been defined, where appropriate, to assess anthropogenic influences. Furthermore, the ICPR has defined target values for almost 60 variables, aiming for compliance with these targets by the year 2000 [ICPR, 1995(a)]. These values, rather than the background concentrations, are used for the assessment of water quality.

Data-reporting obligations, following on from the implementation of EC legislation on the special functions of water used to be fulfilled separately for each directive that contains a reporting obligation. Variables, criteria and reporting formats differed for each of these. Aiming for harmonisation of the reporting obligations, following from the directives dealing with environmental protection in the air, water and waste sectors, a co-ordinating directive of these reporting obligations was agreed to in 1991. This is directive 91/692/EEC for "standardisation and rationalisation of reports concerning the application of certain directives dealing with environmental protection."

Following from this directive, the EC member states have to submit triennial reports for each of the three environmental sectors mentioned above. The first report relates to the years 1993-1995 for the water sector and covers the reporting obligations concerning the directives on dangerous substances (76/464/EEC and subsequent directives), fish water (78/659/EEC), titanium dioxide industry (78/176/EEC). shellfish water (79/923/EEC), ground water (80/68/EEC), surface water for the preparation of drinking water (75/440/EEC and 79/869/EEC) and drinking water (80/778/EEC).

Additional requirements, specifically dealing with the reporting formats ("reporting screens") concerning the questionnaire for the directives for the water sector, have been laid down in directive 92/446/EEC. This directive has recently been amended through directive 95/337/EEC. These amendments, following from obligations for timely submission, will not be put to use until the next three-year annual reporting period (1996-1998).

4. Development of intermediate and ultimate water quality targets

As outlined above, the environmental policy approach for soil and water in the Netherlands is largely reflected in a system of intermediate and ultimate quality targets for water (ground and surface water), sediment and soil(relocation), to be accomplished on a short-to-medium long term time scale. The ultimate target values ("steefwaarden") represent the final goal with respect to environmental quality, whereas the intermediate target values ("grenswaarden") represent an intermediate goal. Periodic adjustment of the intermediate target values should eventually reach the ultimate target values, a process which is called progressive regulation.

4.1 TARGET VARIABLES

The selection of variables for which intermediate and ultimate target values have been established primarily resulted from previous examinations of environmental problems and has been more or less historically determined. With respect to individual substances, a distinction is made between substances which are a direct and actual threat because of their hazardous or other adverse effects on the environment and known occurrence, and substances that might be a threat (known hazardous or otherwise adverse effects, but insufficient information about occurrence). Substances that fit the first description have been designated for routine monitoring (monitoring or "M" list), while substances that fit the second description have been designated for inventory monitoring only (inventory or "I" list).

For the development of target values for soil and water quality, three underlying objectives have been of major importance:

- Risks of damage to the ecosystem have to be negligible;
- Specific functions, of the media to be regulated, must be secured;
- Intermediate and ultimate targets set for different media have to be compatible.

Compliance with the first underlying objective has been achieved through risk assessment studies. Such a risk assessment approach is a key element of the current environmental policy approach in the Netherlands. It seeks for a way of dealing with environmentally hazardous activities that will lead to a uniform protection of the environment from all such activities. The protection levels that should be accomplished determines what should be achieved along the emission based approach.

Two risk levels can be distinguished:

1. The upper limit: the maximum permissible risk level ("MPR")
2. The lower limit: the negligible risk level ("NR").

4.2 TARGET VALUES AND RISK

Target values are linked to such risk levels for man, plants, animals and ecosystem. The maximum permissible risk level is defined, on a media specific and substance by substance basis, as the level that will provide protection to 95% of the species. Chronic toxicity data (No Observable Effect Concentrations: NOEC) for at least four species, from different taxonomical groups, have been used to estimate this maximum permissible risk level.

Given the fact that species are likely to be exposed to more than one substance in real life situations, a safety factor needs to be incorporated to account for the effects that could occur on exposure to more than one substance at a time (e.g. synergetic effects). Bearing this in mind, a safety factor of 100 has been principally applied to deduce the negligible risk levels from the maximum permissible risk levels. Departing from these definitions, ultimate target values have principally been chosen to coincide with the negligible risk levels. Intermediate targets, in turn, never exceed the maximum permissible risk level.

For specific purposes (e.g. soil relocation), additional intervention criteria have been defined at levels higher than the maximum permissible risk level. Such intervention criteria will not be discussed in this paper.

The second underlying objective aims for an inherent compliance with legal provisions made for the protection of water and soil to serve specific functions. It simply requires that ultimate target values never exceed existing criteria that have been implemented for the protection of specific functions of the media regulated.

Finally, the third underlying objective acknowledges the fact that substances in the environment will spread over the various media (ground and surface water and soil and sediment respectively[4])according to a dynamic equilibrium. A specific concentration of a substance in one medium will automatically lead to a related concentration of that substance in another media. This interconnection has been accounted for by linking the target values for a specific substance in different media by means of empirical equilibration constants, bearing in mind that ultimate target values for the individual media should never exceed the negligible risk level.

Legally fixed terms for realising the ultimate target values or the intermediate level have not been set. The aim is to reach the intermediate targets by the year 2000. Furthermore, the aim of Dutch policy is to leave a clean environment for the next generation by the year 2010 which implies that the ultimate target values should be reached in the beginning of the next century. Within current monitoring programs, especially the intermediate targets are being used to assess the environmental condition and to test whether policy approaches aimed at emission reduction are sufficient to assure compliance on a relatively short time scale (by the year 2000).

The target values discussed above relate only to the physico-chemical assessment of water quality. Dutch policy, however, recognises the need for ecological assessment of inland waters in which biological, physical and chemical, as well as environmental and management variables are involved [Klapwijk et al, 1995]. As a completed set of assessment criteria and biological standards for this purpose are not yet available, the "Evaluation Note Water" [V&W, 1994] states that such criteria should be developed for various types of water (i.e. ambient waters, ditches, shallow lakes, canals and sandpits).

In a national programme, set up by the Dutch Foundation for Applied Water Research (STOWA), in which all the water authorities in the Netherlands participate, a reference framework was set up for various sub-types of the main water types described above [STOWA, 1992]. Both biotic and a-biotic factors have been included in these reference frameworks. Quality classes could be developed according to the occurrence in each water type of species characteristic for different stages of degradation. This system is now used to assess ecological quality in Dutch surface waters [CUWVO, 1996].

5. Administrative structures for management, monitoring and assessment

Both licensing and control of discharges and monitoring and assessment of ambient water quality, are highly regionalised in the Netherlands. With respect to surface waters, a distinction can be made between state controlled waters and waters controlled by the regional authorities. The main river arteries such as the Rhine, Meuse, Scheldt and their tributaries, the larger lakes such as the Ijssel lake, and, where appropriate the marine environment (estuaries and coastal waters) are designated "State waters." All other surface waters are controlled by regional authorities [OECD, 1995].

5.1 WATER BOARDS AND OTHER ADMINISTRATIVE BODIES

These regional authorities are mainly the water boards that have been specifically founded to deal with water quantity and quality issues and emissions to surface waters. It is worth mentioning that some of the water boards dealing with water quantity issues date back to the 17th century when they were founded to manage land reclamation activities going on in the Netherlands at that time. In later years water quality and emission issues, originally undertaken by the provinces, were delegated to water boards specifically founded for that purpose. Only the Gronginen and Utrecht provinces still perform such tasks themselves but they are now preparing to delegate these tasks as well. Many small water boards have been incorporated into larger units so that the total number of water boards has decreased from several hundred in the 1960s to about 120.

Depending on the situation of a specific discharge to state controlled waters or to water controlled by the regional authorities, licensing and control of the discharge is performed by the central or by regional authorities. Ten regional directorates of the Directorate General for Public Works and Water Management (RWS) of the Ministry of Transport and Public Works and Water Management (V&W) perform these tasks at the central authority level. Licensing of industrial discharges to sewage systems is the responsibility of the administrators of these sewage systems, generally the municipalities, with the exception of 18 more complex categories of "black list" substances for which the water boards are the authority.

The monitoring of surface water quality for policy preparation and evaluation and testing of compliance with criteria for specific surface water functions are also divided into monitoring of state waters and monitoring of regional waters At the state level monitoring activities are performed by the Institute for Inland Water management and Waste Water Treatment (RIZA), which operates under the Directorate General "RWS" of the Ministry of V&W. At the regional level these tasks are performed by the water boards. Reporting on national water quality is performed by RIZA. For this purpose, a program for data-sharing between the water boards and RIZA has been established. Monitoring activities following from data reporting obligations in the context of international programs and agreements, generally concern state waters. Consequently these activities have been incorporated into the RIZA monitoring program.

Water quality monitoring in coastal waters is performed by a sister institute of RIZA, the National Institute for Coastal and Marine Management (RIKZ). Monitoring of ground-water quality is performed by the National Institute of Public Health and Environmental Protection (RIVM), which operates under the Ministry of Housing, Physical Planning and the Environment (VROM). Co-ordination of environmental monitoring activities is performed by the CCRX (the Co-ordination Commission for the Measurement of Radioactivity and Xenobiotic Substances).

5.2. WATER QUALITY ASSESSMENT

In summary, water quality and/or emission data are used to:

- check compliance with license agreements;
- set emission charges;
- evaluate water quality for the protection of specific surface water functions;
- prepare and evaluate national policy objectives and approaches;
- evaluate the effectiveness of policy installed to meet agreements made within the context of various international programs.

Most of these program specific assessment objectives (information needs) can be achieved by accomplishing some more general primary assessment objectives.

5.2.1. *Primary objectives*

These primary objectives are [RIZA, 1993]

- load calculations and estimates;
- detection of long-term trends in water quality and loads;
- testing compliance with water quality standards and objectives.

Load calculations are of specific importance in the assessment of pollutant balances for sedimentation areas and water bodies that serve as pollutant sinks (e.g. the North Sea). Furthermore, they serve an important purpose in assessing transboundary inflow of pollutants. Information concerning pollutant loads makes it possible to track pollutant sources to location and size, and to evaluate relative contributions from individual pollutant sources to an overall contamination status. Trend detection serves an important purpose in evaluating the effectiveness of instruments installed to achieve policy objectives. Trends, in either loads or concentrations, provide information as to whether such objectives will be reached within the agreed term. Finally, testing water quality for compliance with water quality standards and objectives is the most general assessment objective. It satisfies an important information need for each of the previously mentioned more specific assessment objectives, whenever water quality standards and/or objectives have been set.

On a national basis testing of water quality against intermediate target values (and criteria for specific functions of water) is a major assessment objective in the context of current national environmental policy [CUWVO,1990]. Additionally, trend detection serves to evaluate whether current policy approaches and legal provisions to reduce emission loads from both point and diffuse sources lead to the desired improvement of environmental quality. Trend detection furthermore should confirm whether or not the intermediate target values will be reached within the expected time scale. Load estimates specifically serve to evaluate inputs to sedimentation areas [RIZA, 1993].

With respect to legislative arrangements, assessment of water quality for the protection of specific functions of surface water is mainly analogous to the assessment of water quality against intermediate target values. The water quality criteria for these specific functions serve as the reference framework. Assessment of discharges focuses much more on load assessment, recognising the fact that discharge criteria are generally set for emission loads, with additional concentration criteria where applicable [VROM 1996 and CUWVO, 1996].

5.2.2. *International Agreements*

Within the context of the Rhine Action Program, assessment objectives similar to those for national environmental policy evaluation are present.[ICPR, 1995 and ICPR, 1995(a)]. Even the target values set for water quality are similar to those defined on a national scale. Given the international agreements to reduce emissions of selected pollutants to the Rhine Basin over the period 1985-1995 (by 50 to 70%, depending on the pollutant measured) [ICPR, 1987 and Rhine Ministers Conference, 1989], additional significance is placed on the calculation of output loads from individual tributaries and throughput loads at border crossings. Furthermore, testing of agreements' compliance made within the context of the chloride and chemistry covenants is mentioned as an important assessment objective [ICPR, 1995 and ICPR, 1995(a)].

Within the context of the North Sea Action Program, the assessment objectives established by OSPARCOM Environmental Assessment and Monitoring Committee (ASMO) for the Joint Monitoring Program (JMP) on riverine inputs and direct discharges [OSPARCOM, 1996] largely focus on load calculations. Given the international agreements to reduce pollutant inputs to the North Sea by at least 50% in the 1985-1995 period [North Sea Ministers Conference, 1990] the evaluation of trends in pollution loads (riverine inputs and direct discharges) in particular serves an important additional information need.

Assessment of water quality and emissions to surface water as required by EC legislation aims primarily to check compliance with EC standards, i.e. for emission loads (actual and permitted), and for water quality for specific functions of surface water. Furthermore, aggregate information about geographical locations, local circumstances and legislative provisions needs to be added to the assessment results.

5.3. WATER QUALITY MONITORING

Monitoring strategies largely depend on the assessment objectives that have to be satisfied. Agreements on variables to be monitored and sampling sites, frequencies, and media are generally an important prerequisite to fulfilling assessment objectives. When data obtained by different monitoring institution are to be combined in the assessment phase, it is important to agree on such aspects in advance. The topics on which agreements have been made, as well as the extent to which these agreements are binding, ranging from mandatory to recommended, differ for individual monitoring programmes.

5.3.1. *National monitoring recommendations*

On a national scale, recommendations have been issued for monitoring of substances on the "M-list" (obligatory monitoring list, see above) [CUWVO, 1990]. The recommendations aim to provide a uniform basis for "Monitoring the Water Quality Condition of the Nation" (MWTL), as reported annually (e.g. [CUWVO, 1996]), based on data collected by the water boards and RIZA. Recommendations are included for the selection of:

- sampling sites (sites of special managerial, natural and functional importance, as well as continuity);
- sampling media (in connection with the solubility of individual pollutants);
- minimum sampling frequencies (sufficient statistical significance);
- sampling strategy for sediment (representative information for sedimentation areas);
- sampling and analytical methods (aiming at a possibly high level of standardisation and comparability).

Due to the regionalised nature of the administrative framework for discharge licensing, the diversity in industrial processes and the relatively free format of the licenses, monitoring and data reporting obligations for discharges are not normally uniform [CUWVO, 1996]; rather these issues are tailor-made for each specific discharge. Nevertheless, it is worth mentioning that recommendations, linking sampling frequency to the dimension and fluctuations of the discharge, have been issued by RIZA [Ruiter, 1994]. Furthermore, standardised analytical methods are normally required.

5.3.2. *Harmonising monitoring*

Agreements designed to harmonise monitoring efforts made within the context of the ICPR monitoring programme are of a more stringent nature than those for evaluating water quality on a national scale [ICPR, 1995 and ICPR, 1995(a)]. Apart from agreements on the above mentioned topics, additional emphasis is placed on homogeneity in the cross section of the river and measurement of river flow at the sampling site. Both of these requirements are of special importance when performing load calculations. With respect to analytical methods, the guidelines are less strict in general (only requiring the detection limit to be lower than the target values), but stricter in specific cases (measurements of organic micro-pollutants in water are performed by a single laboratory for all Rhine basin countries).

With respect to the OSPARCOM monitoring programme on riverine inputs and direct discharges [OSPARCOM, 1996], the special emphasis placed on site selection and sampling strategy is worth mentioning. Preferably sites should be selected so as to guarantee a unidirectional freshwater flow. Sampling strategy should concentrate on periods of expected high river flow. Both these recommendations aim at unambiguous calculation of riverine loads to the North Sea.

As with the monitoring programme for the Rhine, special emphasis is placed on measurement of river flow and homogeneity in the cross section at the sampling site. Furthermore, the guidelines of the OSPARCOM Environmental Assessment and Monitoring Committee (ASMO) explicitly state that Quality Assurance (QA) is to be undertaken. Although this requirement is only stated in general terms, it recognises that the comparability of the data within a monitoring programme will benefit from an underlying programme of QA measures.

Finally, the guidelines require that changes in methodology have to be reported. As changes in methodology may lead to (unexpected) changes in monitoring results, this requirement anticipates possible erroneous conclusions as to the occurrence of trends which reflect such changes in methodology.

5.4. DATA SHARING PROTOCOLS

On a national level, the results of water quality assessment and the assessment of discharges to surface water are periodically published in a number of reports. Some of the most important report series are:

- National Water System Report (e.g. [CUWVO, 1996]);
- Environmental Balance (e.g. [RIVM, 1995];
- Emission Data for the Netherlands (e.g. [VROM, 1996]);
- National Water System Exploration (e.g. [Ordelman et. al., 1994]);
- Annual report: Monitoring Results Rhine (e.g. [Heynen et. al. 1991]) and Meuse (e.g. [Boonk et. al., 1991];
- Annual report: Monitoring Results State Waters (e.g. [RIZA, 1991]).

5.4.1. *Integrated view of the National Water System Report*

The first two series provide an integrated view of water quality and overall environmental quality in the Netherlands. The "National Water System Report" includes reviews on ecological water quality in the Netherlands, from 1993 onwards, and the emission status (both from point sources and diffuse sources) in the Netherlands, from 1994 onwards. Reviews on physico-chemical water quality, including sediments and suspended matter, and water quality for specific functions of surface waters, have always been a component in this series. The "Environmental Balance" takes an even wider view, including the sectors air and soil. The "Emission data for the Netherlands" series focuses on emission data only. It covers emissions to both air and water and estimations of emissions from diffuse sources are included.

The "National Water System Exploration" takes a more thematic view of water quality and related issues. Individual reports in this series are generally devoted to a single (group of) substances, or an environmental theme from a sources point of view. The last two series are completely focused on specific monitoring programmes, performed by a single institute (RIZA); i.e. for the Rhine, the Meuse and the State waters overall.

5.4.2. *Monitoring programme data*

Data and information from various monitoring programmes, managed by different authorities, have to be collected and processed especially for the preparation of the first three series. The preparation of the "Environmental Balance" mainly relies on both the "National Water System Report" and the "Emission data for the Netherlands", as well as various other reports, dealing with e.g. the air sector, for this purpose. For both the other two series, data-sharing programmes have been established.

Data on surface water quality are collected on an annual basis by means of a "questionnaire", distributed among the Dutch "water boards". Data collection, by means of this questionnaire is performed by the CUWVO, which is also responsible for publication of the "National Water System Report". Participation in the questionnaire is obligatory. A software version of the questionnaire is available, which allows for (semi) automated data transport from the water boards to the database platform (DONAR) that is use for storing of the data.

Emission data have, until recently, been collected by means of various questionnaires, dispatched by different institutes, among which are the Institute for Technical and Scientific Research (TNO) and the Central Bureau for Statistics (CBS) [VROM, 1996 and Witteveen & Bos, 1995]. Participation in the questionnaire was on a voluntary basis. From 1993 onwards, collection of emission data to surface water is performed by means of a CUWVO questionnaire that is distributed among the authorities responsible for management of emissions to surface water (the water boards).

Apart from national information needs, this questionnaire also covers information requirements that exist on an international level; e.g. as required by EC directive 76/464/EEC and as required within the Rhine - and North Sea Action Programmes. Participation in the questionnaire is obligatory and data collection is closely associated to monitoring obligations, as laid down in the discharge licenses. A software version of the questionnaire will soon be available, allowing for (semi) automated data transport from the water boards to the database platforms that are used for storage of the data. The data are stored in the REVIEW database [Witteveen & Bos, 1995], located at RIZA, and Individual Emission Inventory system [VROM, 1996], located at the Department for Emission Inventory and Information Management at VROM.

5.4.3. *International data sharing*

The international data-sharing programmes in which the Netherlands participates require similar questionnaires to record data-exchange [ICPR, 1995; ICPR. 1995(a); OSPARCOM, 1996 and EC directive 92/446/EEC]. For all of these data reporting obligations, data exchange are now digitised, using software versions of the questionnaires.

Within all of these data-sharing programmes, it is recognised that more than just numbers have to be exchanged in order for an unambiguous evaluation to be performed. Part of this harmonisation is realised by agreements on monitoring activities as discussed above. In addition information about the meaning of the data, must be added within the actual process of data-exchange (and on storage in the databases). Apart from obvious additional information, such as sampling locations, sampling dates and variables, media and units to which the data refer, some less obvious topics might be pointed out based on the reporting formats issued for the various data exchange programmes [CUWVO, 1990; ICPR, 1995; ICPR. 1995(a); OSPARCOM, 1996] considered and personal communication with people involved in these programmes.

5.4.4. *Dealing with detection limit values*

When calculations have to be performed with raw data, detection limit values remain a typical source of error. If detection limit values are not explicitly tagged as such, they may turn up as if they were true measurements, when calculations are performed. If so, they can seriously affect load calculations, and calculations of average and percentile values. Moreover, if detection limits have changed over time they can seriously affect trend detection statistics. Even when detection limits have been tagged, it is of major importance to agree on methods of dealing with them, when performing calculations (e.g. set to zero, or set to half the detection limit). Finally, a clear distinction from missing data needs to be made.

5.4.5. *Calculation methods for delivering aggregated results*

If aggregated results have to be delivered, instead of raw data, it is of major importance that prescribed calculation methods are applied. The results, of load calculations and procedures for testing against target values and criteria, may differ considerably, depending on the calculation method used.

5.4.6. *Particle size distribution for substances associated with suspended matter*

For substances with a low water solubility, measurements are generally performed in suspended matter, or sediment. Concentrations of such substances normally correlate with the small sized particle fraction and organic matter contents of the suspended matter or sediment concerned. It is, therefore, of major importance to add information concerning particle size distribution and organic matter contents to the data, so that standardisation for these parameters can be performed. Furthermore, if such substances are measured in the water phase, we need to determine whether the total or only the dissolved concentration has been determined. Information on suspended matter concentration, as well as particle size distribution and organic matter contents, is also required.

Within this context, it should further be noted that different analytical methods for particle size distribution may lead to incomparable results. Even a single method performed by different laboratories may lead to different results.

5.4.7. *Miscellaneous considerations*

It is important, in addition, to add information on:

- Analytical methods: Apart from the fact that different analytical methods may have different detection limits, systematic errors may lead to different results. It is therefore important to use standardised analytical methods and to perform "round-robin testing".
- Sampling procedure; if different sampling procedures (i.e. grab - versus composite samples) are used by different monitoring institutions, at least the accuracy of the results may be incomparable;
- Adding information, as to whether the data refer to e.g. phosphate-phosphorous or phosphate as such can prevent obvious mistakes in comparing data and performing calculations.

Finally, it was noted that returning data to the original data-source, for validation after storage in a central database, will generally prevent all kinds of errors and obvious mistakes. Within this context, direct personal contact with the experts in charge of monitoring and delivering the data, was seen as an inescapable requirement in data-sharing programmes.

Acknowledgment

The author wishes to thank Rene Breukel, Aukje Hassoldt, Jeanette Plokker, Joost van de Roovaart, Anneke Schäfer and Michael Vossen, at the Institute for Inland Water Management and Waste Water Treatment (RIZA, Lelystad, the Netherlands). for their kind co-operation and information about personal experience in the field of data-sharing.

References

1. Boonk A., Weijden M. v.d., Resultaten van het waterkwaliteitsonderzoek in de Maas in Nederland, *RIZA, Lelystad, the Netherlands*, 1991 (in Dutch).
2. CUWVO (Co-ordination Commission Pollution of Surface Waters Act), National water system report 1994; physico-chemical, functional and ecological water quality and emissions to surface water 1994, *CUWVO working group 7, RIZA, Lelystad, the Netherlands*, 1996 (in Dutch).
3. CUWVO (Co-ordination Commission Pollution of Surface Waters Act), Recommendations for monitoring of substances from the third water management document, *CUWVO working group 5, RIZA, Lelystad, the Netherlands*, 1990 (in Dutch).
4. Heynen R., Weijden M. v.d., Resultaten van het waterkwaliteitsonderzoek in de Rijn in nederland (1971-1990), *RIZA, Lelystad, the Netherlands*, 1991 (in Dutch).
5. ICPR (International Commission for the Protection of the Rhine against pollution), Gesamtmessprogramm fur die Bestandsaufnahme 1995 zur Erfassung der Schadstoffgehalte in den Teilbereichen Wasser, schwebstof, sediment und Organismen, International Commission for the protection of the Rhine against pollution, Koblenz, Germany, 1995 (in German).
6. ICPR (International Commission for the Protection of the Rhine against pollution), Rhine Action Programme, *Ed. Technisches-wissenschaftliches Sekeretariat, Koblenz, Germany*, 1987 (in German).
7. ICPR (International Commission for the Protection of the Rhine against pollution), Vereinbarungen der ICPR fur Messprogramme und Sonderuntersuchungen in den *Teilbereichen Wasser, Schwebstoff, Sediment und Organismen, Koblenz, Germany*, 1995(a) (in German).
8. Klapwijk S.P., Gardeniers J.J.P., Peeters E.T.H.M. and Roos C., Ecological Assessment of Water Systems, In Monitoring Tailor-made, An international workshop on monitoring and assessment in water management, 20-23 September 1994, Beekbergen, the Netherlands, *Proceedings of the workshop*, RIZA, Lelystad, the Netherlands, 1995.
9. North Sea Ministers Conference, Final declaration of *the Third International Conference on the Protection of the North Sea*, The Hague, 1990.

10. Ordelman H.G.K., Noort P.C.M. van, Hulscher T.E.M., Beek M.A., Steenwijk J.M. van, Frintrop P.C.M. and Evers E.H.G., Watersysteemverkenning 1996; een analyse van de problematiek in het aquatisch milieu: Organofosforbestrijdingsmiddelen, *RIZA and RIKZ*, 1994 (in Dutch).
11. OSPARCOM (Oslo and Paris Commissions), Principles and methodology of the Joint Monitoring Programme, Paris, France, 1990.
12. OSPARCOM (Oslo and Paris Commissions), Summary record of the meeting of ASMO, Vila Franca do Campo, 4-8 March 1996, *OSPARCOM Environmental Assessment and Monitoring Committee (ASMO)*, Paris, France, 1996.
13. Rhine Ministers Conference, *Kommunique über die 10 Ministerkonferenz zum Schutze des Rheins gegen Verunreinigung*, Brussels, 30 November 1989.
14. RIVM (National Institute for Public health and Environmental Protection), Environmental Balance 1995, Bilthoven, the Netherlands, 1995.
15. RIZA (Institute for Inland Water Management and Waste Water Treatment), *Yearbook*, Lelystad, the Netherlands, 1991 (in Dutch).
16. RIZA, Adriaanse M., Optimalisatie routinematig onderzoek waterkwaliteit rijksbinnenwateren, *RIZA, Lelystad, the Netherlands*, 1993 (in Dutch).
17. Ruiter, T. de, Bepaling meetfrequentie ter vaststelling van de vervuilingswaarde van water (werkdocument), *Ministry of V&W, RIZA*, 1994 (in Dutch).
18. State Journal, *The Pollution of Surface Waters Act*, SDU, The Hague, 1992, 628 (in Dutch).
19. State Journal, *Decision on Quality Targets and Monitoring of Surface Waters*, SDU, The Hague, 1983, 606 (in Dutch).
20. State Journal, *Environmentally Hazardous Substances Act*, SDU, The Hague, 1992, 632 (in Dutch).
21. State Journal, *Hygiene and Safety of Swimming Waters Act*, SDU, The Hague, 1982, 494 (in Dutch).
22. State Journal, *Mains Water Supply Act*, SDU, The Hague, 1984, 220 (in Dutch).
23. State Journal, *Soil Protections Act*, SDU, The Hague, 1986, 374 (in Dutch).
24. State Journal, *Groundwater Act*, SDU, The Hague, 1981, 392 (in Dutch).
25. STOWA (the Dutch Foundation for Applied Water Research), *Ecologische beoordeling en beheer van oppervalktewater; wetenschappelijke verantwoording van het beoordelingssysteem voor stromende wateren*, STOWA, Utrecht, the Netherlands, 1992 (in Dutch).
26. VROM (Ministry of Housing, Physical Planning and the Environment), V&W (Ministry of Transport and Public Works and Water Management), EZ (Ministry of Economic Affairs), LNV (Ministry of Agriculture, Nature and Fisheries), *National Environmental Policy Plan: To choose or to lose*, SDU, The Hague, the Netherlands, 1989.
27. VROM (Ministry of Housing, Physical Planning and the Environment), *Essential Environmental Information: the Netherlands*, The Hague, the Netherlands. 1991.
28. VROM (Ministry of Housing, Physical Planning and the Environment), *Environmental Quality Targets for Soil and Water*, Dutch Parliament, 1990-1991, 21 990, No 1, SDU, The Hague, 1991(a) (in Dutch).
29. VROM (Ministry of Housing, Physical Planning and the Environment), *Emission data for the Netherlands; 1993 and estimates for 1994*, Publication series Emission

Inventory, Inspectorate General for Environmental Protection, Department for Emission Inventory and Information Management, the Hague, the Netherlands, No 30, 1996.

30. V&W (Ministry of Transport and Public Works and Water Management), *Third Water Management Document (water voor nu en later)*, Dutch Parliament, 1988-1989, 21 250, No 1-2, SDU, The Hague, 1989 (in Dutch).
31. V&W (Ministry of Transport and Public Works and Water Management), *Evaluation Note Water, Dutch Parliament*, 1993-1994, 21 250, No 27, SDU, The Hague, 1994 (in Dutch).
32. Witteveen & Bos, *Aspectrapport Emissies*, prepared for RIZA, Lelystad, the Netherlands, 1995 (in Dutch).

---

[1] The Groundwater Act [State Journal, 1981, 392] only covers abstraction and artifical recharge of groundwater management.

[2] More modifications to this framework are anticipated in the near future. The "Fourth Water Management Document," to be agreed on by the end of 1997, is expected to put still more emphasis on an integrated approach towards environmental problems (accountring for economical, ecological and sociological developments). Following from this, changes in the requirements for data collection can be anticipated.

[3] Similar programs for the Meuse and the Scheldt are in preparation

[4] Where appropriate, efforts to account for equilibration to air as well are expected for future refinements.

WATER QUALITY MANAGEMENT PROGRAMMES IN THE UNITED KINGDOM

E. G. BELLINGER
Central European University
Budapest, Hungary

KEYWORDS / ABSTRACT: water quality programmes / United Kingdom water quality monitoring

The UK has a long history of dealing with water pollution problems; successive governments having initiated and modified the regulatory framework frequently over time. Currently, the UK is subject to Europe Union Council Directives on water. Unlike other member states, the UK has no river crossing national boundaries. Its water quality standards and the rationale behind them are discussed and UK regulations described. A new regulatory organisation, the Environmental Agency, was formed in March of 1996. In 1989 the largely public water supply and regulatory bodies were replaced by a privatised industry and a separate, independent regulatory authority. These and other changes are explored with reference to monitoring and the exchange of data. These are areas where a considerable amount of work is needed on cross-calibration and harmonisation.

1. Introduction

Any consideration of United Kingdom (UK) water regulations must take into account European Union (EU) requirements. EU environmental requirements are imposed on member states principally in the form of directives, which are binding as to the results to be achieved but allow each member state the choice of methods for doing so. A number of EC directives and UK laws concerning the aquatic environment are mentioned in this paper. These are listed in appendices I & II.

Water pollution was one of the first areas addressed by EU environmental policy initiatives. The earliest strategies for combating it dated back to the first Environmental Action Programme of 1973. Early strategies to control water quality focused on establishing Environmental Quality Standards (EQS). Standards were set for waters used for specific purposes such as surface waters for drinking (75/440), bathing water (76/160), shellfish waters (79/923) and freshwater fish (78/659). Member states were to apply and set their own limit values, although the Directive laid down minimum standards. Time limits were set for implementation. Because of the

I. Lyons Murphy (ed.), Protecting Danube River Basin Resources, 81–88.

large degree of flexibility permitted, difficulties with implementation have been experienced.

The need to control the discharge of dangerous substances to water resulted in the so-called Dangerous Substances Directive (76/464). The Commission supported a directive to control pollution by establishing uniform emission limits. The UK disagreed with this supporting instead emission limits which would set specified environmental quality in the receiving water. This was described as the Environmental Quality Objective (EQO) approach. The main reason for disagreement was the related geography - the UK is an island surrounded by tidal waters and has no international rivers.

The EQO approach looks at an overall environmental picture but would also have administrative difficulties of implementation in certain countries. Eventually a compromise was reached and various Directives were added to the original 76/464 with either limit values or quality objectives for "black list" substances provided and the use of EQO for "grey list" substances. The split has led to difficulties in establishing further controls because of the lack of equivalence between the two approaches.

In addition to these two policies a pollution source approach was also used which targeted specific industrial sectors. The first two were the Titanium Dioxide and the pulp and paper industries although proposals for the latter have now been withdrawn. The Titanium Dioxide Directive was the start of the Integrated Pollution Control approach recently culminating in the Directive on Integrative Pollution Prevention and Control (COM(93)423) which aims to control discharges to all media i.e. air, land and water.

The Fourth Action Programme on the Environment (1987-92) put in place the Commission's desire to control freshwater and marine pollution from both point and diffuse sources. This initiated a new approach to water pollution which resulted in Directives on Nitrates from Diffuse sources and the Directive on Urban Waste Water Treatment (UWWT). The formed encourages farmers to use less fertiliser and engage in less polluting farming activities whilst the UWWT is concerned with the collection, treatment and discharge of urban waste water and waste waters from specific industries.

2. The approach in England and Wales

In 1989 there was a major restricting of the UK water industry. The previous, largely public, utilities which were concerned with providing drinking water, the treatment and disposal of liquid wastes and the regulation of rivers (including flood control) were replaced by a privatised industry and a separate, independent regulatory authority. Under the previous arrangements the Regional Water Authorities could be ascribed as both "Poacher and Gamekeeper" i.e. they were at the same time both a major polluter (through unsatisfactory sewage treatment works discharges) and the regulating authority. Few prosecutions occurred during this period against themselves! This situation was obviously not entirely satisfactory although the concept of integrating all water activities within the catchment of a river or lake under one umbrella had some merit. The 1989 Water Act has now changed this and has been supplemented by the

Environmental Protection Act (1990) and re-enacted as the Water Resources Act (1991).

3. Implementation of water regulations in the UK

At the present time the regulations regarding water in the UK are embodied in the Water Act (1989), the Water Resources Act (1991) and the Environmental Protection Act (1990). Although the Water Act (1989) set down a system for Statuary Water Quality Objectives (SWQO) many aspects concerning water pollution need further consolidation under the WRA 1991 which dealt with 9 separate aspects: the National Rivers Authority (NRA), Water Resources Management, Control of Pollution of Water Resources, Flood Defence, Fisheries, NRA Finance, Land and Works, Provision of Information and Miscellaneous.

Up until this year there was a number of environmental agencies in England. These were the NRA, Secretary of State for the Environment, HMIP, Water Utility companies, DG Water Services, MAFF, EC Commission. From the start of this financial year (April 1996) a new Environmental Agency will combine many of these roles.

The NRA was the most important agency for regulating the quality of water. With the use of SWQO its aim was to improve water quality step by step over a number of years. This would be done with a view to certain water quality criteria which would allow classification according to the following criteria:

a) the general requirements as to the purpose for which the receiving waters were to be suitable;
b) specific requirements as to substances that are to be present in or absent from the waters and as to concentrations of substances required to be present.

These water quality objectives would be specific for each catchment and are not necessarily nation-wide.

The NRA was provided with powers to establish a licensing scheme for discharge in order to achieve the SWQOs and also to be able to bring criminal proceedings against dischargers not meeting their authorised requirements. In order to comply with EC requirements the UK needed to introduce a statutory classification system for waters used for specific purposes as well as the SWQOs and water quality standards.

3.1. CLASSIFICATION SYSTEM

The provision of a classification system was empowered under WRA91 and could cover all controlled waters i.e. territorial waters, coastal waters, inland freshwater and ground water's. The concept was to base the classification on the purpose for which the water was to be used and the presence or absence of certain substances i.e. upon water chemistry. Whilst exact criteria for the classification were not specified in practice these included:

- suitability for specific use of water
- concentrations of specified substances
- any other relevant characteristics
- any combination of the above

In England and Wales four areas of Classification Regulations have been specified:

- Surface Waters (Classification) Regulations 1989 (SI 1989 No 1148)
- Surface Waters (Dangerous Substances) (Classification) Regulations 1989 (SI 1989 No 2286)
- Surface Waters (Dangerous Substances) (Classification) Regulations 1992 (SI 1992 No 337)
- Bathing Waters (Classification) Regulations 1991 (SI 1991 No 1597)

In Scotland two areas of classification have been specified:

- Surface Waters (Classification) (Scotland) Regulations 1990 (SI 1990 No 121)
- Surface Waters (Dangerous Substances) (Scotland) Regulations 1990 (SI 1990 No 1026)

The Secretary of State set SWQO's on a catchment by catchment basis on the advice of the then NRA (now Environmental Agency) and to be implemented by them. A water quality objective is thus set for the prescribed water and a classification given together with a date for achieving that classification. The original aim was to cover all waters by 1995 but this did not happen.

Any SWQO will consist of several elements. It will have an exact location (stretch of river covered), will indicate the water-class or classes (fisheries, abstraction, water sports and conservation) and thus indicate various water quality standards for such things as oxygen, ammonia and BOD. The SQUOs will be the goal driving the levels set for consent for discharges and are the main pollution tool. The NRA has been the authority in England and Wales for ensuring compliance with EC Directives and it had the power to institute criminal proceedings against anyone polluting a controlled water as well as being able to enforce regulations initiating measures to prevent the pollution. Protection zones could also be designated for ground water restricting certain activities within those zones. Discharges could also be prohibited especially if it contained a prescribed substance which in fact would be a criminal offence.

Consents for discharge (authorised by the NRA, now EA, under part III of WRA 91) are issued on an individual basis and can be subject to special conditions e.g. temperature, volume, outlet location etc. In the event of disagreement about these an appeal can be made to the Secretary of State. Before a consent can be granted it must be published and interested parties given 21 days in which to object.

3.2. INFORMATION AVAILABLE UNDER ALL PROVISIONS

The Environmental Protection Act 1990 (Part I) introduced some further changes to water pollution control by making the most dangerous discharges (Red List Substances) subject to prior approval by HMIP under a system of Integrated Pollution Control (IPC). Embodied in all of these provisions is a requirement to monitor and make available information. The information may not necessarily be published but should in some form be accessible to the public. Discharge consent conditions, for example, are not published as such but are available for consultation at the regional offices of the NRA (EA). Monitoring of the discharges is often the responsibility of the discharger together with the need to keep records which are available to the NRA (EA) whose inspectors have a right to monitor an effluent themselves as a check but this information is not made public.

There is a requirement in the EC Directive on Drinking Water to monitor in order to assess whether standards are being maintained (Water Supply (Water Quality) Regulations 1989 amended 1991). This details the number of samples to be taken, parameters to be analysed and the frequency of analysis. Where no standard frequency is set the supplier must take samples as they believe necessary. There are also requirements for taking, handling and storing samples. This scheme is monitored by the Drinking Water Directorate (DWI). Currently about 3.6 million separate analytical determinations are done each year on drinking water by the suppliers in England and Wales. Access to this information is defined under the DoE/Welsh Office Water Supply (Water Quality) Regulations Statutory Instrument (1991) No 1837 and Guidance on Safeguarding the Quality of Public Water Supplies (1989). These require results to be kept for public record to which consumers can have access and take copies.

3.3. THE NRA GENERAL WATER QUALITY CLASSIFICATION SCHEME

Surface waters are classified according to various classes (derived from the original National Water Council Scheme in the 1970s). The current scheme is based on a number of so-called windows through which water quality can be viewed. The first of these windows is based on water chemistry. Other windows will be added to the scheme based on biology, nutrients and aesthetic quality. The river classes based on the chemical window are illustrated below and use the following parameters: BOD, dissolved oxygen and ammonia. The percentile compliance required is also given. The required standards for the above parameters are also given.

It is essential in any such scheme to harmonise both the sampling procedures and the analytical techniques used or at least the detection limits etc. required by a chosen technique. Unless this is done intercomparability of results cannot be guaranteed.

When the monitoring programme is extended to include biological, nutrient and aesthetic parameters the problem of intercomparability become even more of a problem. Each approach will measure different things with different objectives in mind; biological to look at water quality as it affects organisms, nutrients to assess eutrophication and aesthetics to consider leisure and sport activities. Whilst each approach will have value much research is needed to examine the differences in interpretation of quality between each approach, what gives rise to these differences and how some equivalence between them may be arrived at. There are also problems in deciding on which technique to use within individual approaches. For example in biological monitoring do you use the Saprobien System, the UK Biological Monitoring Working Party (BMWP) or RIVPACS or some other? For nutrient assessment does one measure individual compounds or total amounts? For aesthetics it becomes potentially very subjective. These are areas where a considerable amount of work is needed on cross-calibration and harmonisation.

4. Conclusions

There is a general need to monitor water courses for their quality in order to ascertain what and how improvements may be made so that the resource may be enhanced. The perception of quality will vary depending upon the end use and this must be specified at the outset by clearly stating the objectives of the exercise. Once the objectives have been defined strategies must be agreed which include details of sampling (including frequency) and analyses. The information gathered must be made freely available to all interested parties so that meaningful discussions and participation can take place in working for overall improvements. In order to achieve these improvements some mechanism for policing any agreed standards must exist and this must have teeth enough to work.

It is essential in any such scheme to harmonise both the sampling procedures and the analytical techniques used or at least the detection limits etc. required by a chosen technique. Unless this is done intercomparability of results cannot be guaranteed.

When the monitoring programme is extended to include biological, nutrient and aesthetic parameters the problem of intercomparability become even more of a problem. Each approach will measure different things with different objectives in mind; biological to look at water quality as it affects organisms, nutrients to assess eutrophication and aesthetics to consider leisure and sport activities. Whilst each approach will have value much research is needed to examine the differences in interpretation of quality between each approach, what gives rise to these differences and how some equivalence between them may be arrived at. There are also problems in deciding on which technique to use within individual approaches. For example in biological monitoring do you use the Saprobien System, the UK Biological Monitoring Working Party (BMWP) or RIVPACS or some other? For nutrient assessment does one measure individual compounds or total amounts? For aesthetics it becomes potentially very subjective. These are areas where a considerable amount of work is needed on cross-calibration and harmonisation.

Appendix 1. Main EC Water Directives

Subject	Directive Number and Title Environmental Quality Standards	Official Journal Reference
Water Standards for freshwater fish	78/569 Directive on the quality of fresh waters needing protection or improvement in order to support fish life	OJ L222, 14.8.78
Shellfish Waters	79/923 Directive on the quality required for shellfish waters	OJ L281, 10.11.79
Bathing Water	76/160 Directive concerning the quality of bathing water	OJ L31, 5.2.76
Surface water for drinking	75/440 Directive concerning the quality required for drinking water in the Member States	OJ L194, 25.7.75
	79/869 Directive concerning the methods of measurement and frequency of sampling and analysis of surface water intended for the abstraction of drinking water in the Member States	
Titanium Dioxide	78/176 Directive on waste from the titanium dioxide industry	OJ L54, 25.2.78
Dangerous Substances	76/464 Directive on pollution cause by certain dangerous substances discharged into the aquatic environment of the Community (Framework Directive)	OJ L129, 18.5.76
Nitrates from Agricultural Sources	91/271 Directive concerning urban waste water treatment	OJ L375, 31.12.91
Groundwater Protection	80/68 Directive on the protection of groundwater against pollution caused by certain dangerous substances	OJ L20, 26.1.80
Urban Waste Water	91/676 Directive concerning the protection of waters against Pollution	OJ L135, 30.5.91

Appendix II. Main UK Legislation concerning water

Subject	Act	Reference
Water industry and privatisation and formation of NRA	Water Act 1989	
Protection of the Environment	Environmental Protection Act 1990	
Water Resources	Water Resources Act 1991	
Water Classification	Surface Waters (Classification) Regulations 1989	SI 1989 No 1148
Water Classification	SW (Dangerous Substances) (Classification) Regulation 1989	SI 1989 No 2286
Water Classification	SW (DS) (Class) Regs 1992	SI 1992 No 337
Water Classification	Bathing Water (Classification) Regulations 1991	SI 1991 No 1597

GQA chemical grading for rivers and canals

Water Quality	Grade	Dissolved Oxygen	Biochemical Oxygen Demand (ATU1)	Ammonia
		(% saturation) 10-percentile	(mg/l) 90-percentile	(mgN/l) 90-percentile
Good	A	80	2.5	0.25
	B	70	4	0.6
Fair	C	60	6	1.3
	D	50	8	2.5
Poor	E	20	15	9.0
Bad	F2			

1 as suppressed by adding allyl thio-urea

2 i.e. quality which does not meet the requirements of grade E in respect of one or more determinants.

DESIGN AND IMPLEMENTATION OF THE NATIONAL WATER-QUALITY ASSESSMENT PROGRAM – A UNITED STATES EXAMPLE

Understanding the Limitations of Using Compliance-Monitoring Data to Assess the Water Quality of a Large River Basin

DAVID J. WANGSNESS
United States Geological Survey
Peachtree Business Center
3039 Amwiler Road, Suite 130
Atlanta, Georgia, USA 30360-2824

KEY WORDS / ABSTRACT: United States Geological Survey / National Water-Quality Assessment / water quality monitoring / water quality data-sharing

In the 1980s it was determined that existing ambient and compliance-monitoring data could not satisfactorily evaluate the results of hundreds of billions of dollars spent for water-pollution abatement in the United States. At the request of the US Congress, a new programme, the National Water-Quality Assessment, was designed and implemented by government agency, the US Geological Survey (USGS). The Assessment has reported status and trends in surface- and ground-water quality at national, regional, and local scales since 1991. The legislative basis for US monitoring and data-sharing policies are identified as well as the successive phases of the design and implementation of the USGS Assessment. Application to the Danube Basin is suggested. Much of the water-quality monitoring conducted in the United States is designed to comply with Federal and State laws mandated primarily by the Clean Water Act of 1987 and the Safe Drinking Water Act of 1986.

Monitoring programs generally focus on rivers upstream and downstream of point-source discharges and at water-supply intakes. Few data are available for aquifer systems, and chemical analyses are often limited to those constituents required by law. In most cases, the majority of the available chemical and streamflow data have provided the information necessary to meet the objectives of the compliance-monitoring programs, but do not necessarily provide the information required for basin-wide assessments of water quality at the local, regional, or national scale.

1. Introduction

During the period 1972-86 an estimated $541 billion was expended for water-pollution abatement in the United States, and in 1986 the US Congress asked if the Nation's

I. Lyons Murphy (ed.), Protecting Danube River Basin Resources, 89–103.

water quality was improving as a result of those expenditures. For various reasons, this basic question could not be answered satisfactorily using only existing ambient- and compliance-monitoring data. Therefore, the US Geological Survey (USGS) was requested to design and implement a National Water-Quality Assessment (NAWQA) Program to address the questions related to status and trends in surface- and ground-water quality at national, regional, and local scales. The program was fully implemented in 1991.

An initial task of the NAWQA Program was to locate and evaluate existing data, and use those data to help meet as many of the program goals as possible. In general, these data are stored in computerized data bases or as paper files and are available upon request. The sharing of these data generally is not an issue within and among the various governmental agencies. Far greater issues are standardization, quality assurance, and some assurance that the shared data meet program goals and help to answer the questions being asked.

Because the existing data were not adequate to meet the goals of the NAWQA Program, the USGS has designed, and is implementing, water-quality monitoring networks within large river basins and major aquifer systems that, when combined with existing programs, will provide the information necessary to meet program goals. Great care has been taken to use standardized study approaches and techniques for data collection, laboratory analyses, and quality assurance so that information is comparable throughout the United States and can be compiled at various scales.

Some of the experiences gained by the USGS during the development of the NAWQA Program may be useful when developing an international data-sharing programme for the effective management of Danube River basin resources. It is important for participating nations to agree upon ways to share data, but if those data were collected as part of compliance-monitoring programs, similar to those in the United States, it cannot be assumed that those data alone will provide the information needed to assess the quality of the Danube River, or design effective management programmes. In addition to agreeing to share data, it is equally important to agree upon mutual goals and a valid design for the assessment programme.

It is important to start by identifying questions and issues related to the water quality of a river basin, prioritize those questions and issues, and based on that prioritization, identify and prioritize a list of data and information needs. The next step is to identify and locate sources of existing data; consolidate the data; process, summarize, and evaluate the data; and share the results. This information could then be used to determine which of the priority questions could be answered using existing data, which could not be answered, and what data are necessary to fill information gaps. An assessment programme could then be designed to fill the gaps, taking care to standardize approaches, techniques, and data bases, and to use new data in conjunction with existing data to address the priority questions. Standardization is critical throughout the programme so that the combined efforts of all participants will provide the information needed to answer the questions and, thereby, attain the agreed upon goals.

This evaluation is not meant to be a criticism of the available data from compliance-monitoring programs; but rather to point out that those data were collected to meet specific goals and to answer specific, short-term questions, and not for the purpose of meeting the goals of a long-term, river-basin-scale assessment. All data sets can be an important source of information for a water-quality assessment program, but their limitations need to be recognized and evaluated, and often, new data need to be collected to address specific assessment goals.

2. Legislative Background

2.1. WATER-QUALITY LEGISLATION

Federal water-quality legislation in the United States began with the Federal Water Pollution Control Act as amended by the Federal Water Pollution Control Act Amendments of 1972 (PL 92-500), and further amended by the Clean Water Act of 1977 (PL 95-217) and the Clean Water Act Amendments of 1978 (PL 95-676). Current legislation is derived from the following:

Clean Water Act (PL-100-4)	1987
Safe Drinking Water Act (PL-99-339)	1986
Superfund Amendments (PL-99-499)	1986
Resource Conservation and Recovery Act (PL-94-580)	1976

Also included are a wide variety of other Federal and State mandates [1]. It is also important to note that the Clean Water Act presently (1996) is in the US Congress for reauthorization.

Water-quality legislation in the United States provides the basis for the development and implementation of standards and guidelines. It also designates the US Environmental Protection Agency (USEPA) as the regulatory authority at the Federal level. In most cases this authority is delegated to state agencies that are responsible for environmental protection, provided that the state's mandates meet or exceed the Federal mandates.

National standards and guidelines for water quality are established for the United States by agencies of the US Government or by the National Academy of Sciences (NAS) and National Academy of Engineering (NAE). Standards and guidelines are specific to one sampling medium (water, bed sediment, and fish and shellfish tissue) and focus on protection of one or more beneficial uses of the hydrologic system (drinking water, fish and shellfish consumption, aquatic organisms, and wildlife). The term "standard" refers to threshold values that are legally enforceable by the USEPA and/or designated state regulatory agencies. In water, USEPA maximum contaminant levels (MCLs) for drinking water, and levels of chemicals, such as pesticides, in edible fish and shellfish tissues are enforceable. The term "guideline" refers to threshold values that have no regulatory status but are issued in an advisory capacity. For instance, NAS/NAE may recommend maximum concentrations in water for protection of aquatic life that often become precursors to USEPA aquatic-life criteria or guidelines, but are not considered standards until they have undergone notice and comment procedures under the Administrative Procedures Act. Many states use federal guidelines as the basis for state standards.

The USGS is the Federal government's earth-science agency and has no regulatory authority; however, the agency does conduct studies that are jointly funded by various levels of government. These studies provide data that are used to directly support federal and state requirements, as well as providing interpretive information that helps water-resource managers understand their hydrologic systems. Standards and guidelines are useful tools in water-quality studies, since they provide a measure for comparison. However, valid use of a given standard or guideline requires an understanding of its technical basis and underlying assumptions. Several detailed references describe the technical basis and underlying assumptions, and provide lists of current standards and guidelines (see Camp Dresser & McKee, Inc. [2]; NAS/NAE [3]; Nowell and Resek [4]; USEPA [5], [6], and [7]). Updated information is available at the USEPA home page on the World Wide Web at http://www.epa.gov.

2.2. WATER-QUALITY MONITORING PROGRAMS

One of the greatest challenges faced by water-resource scientists is acquiring reliable data and information that will guide the use and protection of the Nation's water resources. That challenge is being addressed by many local, State, and Federal agencies and academic institutions in the United States that collect and store water-quality data as a part of routine compliance-monitoring, water-quality assessment or ambient monitoring programs, and research activities in response to federal and state laws mandated primarily by the Clean Water Act of 1987 and the Safe Drinking Water Act of 1986. An additional use of water-quality data and information is to provide a basis on which regional and national-level policy decisions can be based. Informed decisions must be based on sound information. As a society we need to know whether certain types of water-quality problems are isolated or ubiquitous, whether there are significant differences in conditions among regions, whether the conditions are changing over time, and why these conditions change from place to place and over time. The information can be used to help determine the effectiveness of existing water-quality policies and to help analysts determine the need for and likely consequences of new policies.

Much of the surface-water monitoring conducted in the United States is compliance monitoring and focuses on chemical concentrations and flows discharged from point sources, locations upstream and downstream of those sources, and water-supply intakes (see ITFM [1] and Powell [8]). For example, the State of Wisconsin estimates that more than 98 percent of it's monitoring program is compliance monitoring, and Washington State estimates 80-90 percent of its monitoring program to be compliance monitoring. The goal of compliance-monitoring programs is to determine whether or not a given discharger, such as a municipal wastewater treatment facility, is in compliance with its discharge permit. Water samples are periodically collected from both the effluent and the receiving water body, and are analyzed for concentration of selected constituents to determine if the concentration exceeds a water-quality standard or limit defined in the discharge permit. A large percentage of the water-quality data in the United States has been collected as a part of compliance monitoring programs.

Ambient-monitoring and water-quality assessment programs generally focus on issues related to water-resource management of a river basin for multiple uses. There generally are multiple goals for these programs, such as: 1) describing the current status and historical trends in water quality, 2) comparing and contrasting the effects of point-source and nonpoint-source inputs, and 3) providing clear and concise summaries of sound scientific interpretations and recommendations to land- and water-resource managers. A small percentage of the water-quality data in the United States have been collected as a part of ambient monitoring programs or river-basin assessments. A small percentage of the water-quality data in the United States have been collected as a part of ambient monitoring programs or river-basin assessments.

In most cases, the majority of the available chemical and flow data have provided the information necessary to meet the objectives of the compliance-monitoring programs, but do not necessarily provide the information required for basin-wide assessments of water quality at the local, regional, or national scale. Therefore, the majority of existing data have provided some of the information necessary to satisfactorily support some of the short- and long-term decisions concerning water-resources management in the United States.

3. The National Water-Quality Assessment Program

Following the appropriation of funds by Congress in 1986, the USGS began a pilot program in seven project areas to develop and refine a plan for the National Water-Quality Assessment (NAWQA) Program [9]. In 1991, the USGS began full implementation of the program [10]. The NAWQA program builds upon an existing base of water-quality studies of the USGS, as well as those of other Federal, State, and local agencies. The objectives of the NAWQA program are to:

1. Describe current water-quality conditions for a large part of the Nation's freshwater streams, rivers, and aquifers.
2. Describe how water quality is changing over time.
3. Improve understanding of the primary natural and human factors that affect water-quality conditions.

This information will help support the development and evaluation of management, regulatory, and monitoring decisions by other Federal, State, and local agencies to protect, use, and enhance water resources.

The goals of the NAWQA Program are being achieved through ongoing and proposed investigations of 60 of the Nation's most important river basins and aquifer systems (referred to as study units). These study units are distributed throughout the Nation and cover a diversity of hydrogeologic settings. More than two-thirds of the Nation's freshwater use occurs within these study units and more than two-thirds of the people served by public water-supply systems live within their boundaries. The 60 study units have been divided into three groups of 20 study units each, and the intensive data-collection phases have been staggered to allow efficient and effective use of funds and human resources. The first 20 studies began in 1991, the second group began in 1994, and the third group is planned to begin in 1997. The first group of 20 study units is planned to begin a second cycle of study in the year 2000, and the cycle is intended to continue into the future, so as to provide both short-term information necessary for today's water-resource management decisions, and the long-term information needed for policy decisions.

3.1. THE NAWQA PROGRAM DESIGN

Study-unit investigations are conducted at the river-basin or aquifer-system scale, and will comprise the foundation on which national and regional assessments will be based. National and regional synthesis of data, based on analyses of an aggregation of comparable information obtained from the study units, is a major component of the program [10]. This effort focuses on selected water-quality topics using nationally consistent information. Comparative studies will explain differences and similarities in observed water-quality conditions among study areas and will identify changes and trends and their causes. The first topics addressed by the national synthesis are pesticides and the nitrogen and phosphorus nutrients, followed by volatile organic compounds, and aquatic biology as additional topics. Discussions of these and other water-quality topics will be published in periodic summaries of the quality of the Nation's ground and surface water as the information becomes available.

Study-unit investigations consist of four phases. The first phase, *planning and study design*, and the second phase, *analysis of existing data,* have considerable overlap. The third phase, the implementation of the study design, is a *3-year period of intensive data collection*. The fourth phase is a period of *intensive data analysis, interpretation, and report production*; coincident with *low-intensity monitoring* meant to provide a link between high-intensity periods. Throughout all phases of the study, and at all levels of the program, emphasis is placed on using nationally consistent study designs and approaches, and sampling and analytical techniques to provide nationally consistent information that will address a pertinent set of questions. The approaches being used by all study-unit teams include conducting retrospective analyses of existing data; establishing a nationwide, long-term-monitoring network designed to assess existing water-quality conditions and provide a data base for trend analyses; and conducting process-oriented studies designed to provide a better understanding of the relation between land- and water-use activities and water-quality conditions.

3.1.1. *Planning and Study Design*

Planning and study design was the initial task for the 20 study-unit teams that were formed in 1991. This was an iterative process that included addressing local issues, and at the same time assuring that the information provided by each study unit could ultimately be aggregated to address issues at the regional and national scale. To assure compatibility at all scales, standard lists of constituents were chosen, study approaches were designed, protocols for collecting and processing samples were developed, a quality-assurance plan was designed and implemented, and it was decided that all samples would be analyzed by one centralized laboratory using standardized analytical techniques. Compatible computerized data bases and geographic information systems (GIS) also were designed to be a part of a distributed-information system. To assure local involvement, liaison committees comprised of representatives from local, state, and Federal government agencies, environmental groups, and universities were formed to provide a routine exchange of ideas and information.

3.1.2. *Analysis of Existing Data*

The second phase for each study unit has been the analysis of existing data. In most study units much of the data for retrospective analyses were obtained from the USEPA's Water Storage and Retrieval (STORET) data base and from the USGS's National Water Information System (NWIS) data base. The STORET data base is available to all data-collection agencies for the storage and retrieval of those agencies' data, as well as the data of others. However, USEPA relies on each contributing agency to enter and quality assure their own data. Agency-specific data bases are not readily available, or are not yet automated, so STORET remains the best source for water-quality data. The NWIS data base is managed by the USGS and data inputs are strictly quality assured. The data are public information and available to users upon request, but direct access to the data base is limited. Private industries also monitor water quality but not all their data are available; in particular, the ground-water quality data.

Each study-unit team compiled and analyzed existing water-quality data from water-resource agencies at all governmental levels. Those analyses determined the extent to which NAWQA goals could be met using existing data, and pointed out issues and areas that needed additional study. Because the types, amounts, and quality of data varied between study units, the retrospective analyses were a critical part of the iterative design process for the NAWQA Program as each study unit developed a plan to fill data gaps in a nationally consistent manner.

To meet the goals of assessing the Nation's water quality, the following were needed:

1. long-term monitoring sites spatially distributed throughout a river basin and representative of a composite of upstream inputs;
2. long-term data sets that included a broad list of constituents, including stream flow, that were temporally distributed throughout a year and were representative of all hydrologic conditions, including extremes such as droughts and floods;
3. computerized data sets available in standard and easily readable formats;
4. computerized data sets that were complete and accurate -- that is, a minimum standard set of data fields having data that were quality assured (checked and verified);
5. assurance that the same or similar sample-collection techniques and laboratory analytical techniques were used, and that all techniques were well documented; and
6. computerized data sets containing ancillary data that could be used to help explain water-quality information.

In general, the following were available:

1. compliance-monitoring sites that often were poorly distributed spatially and heavily influenced by a point-source inputs and regulated flows;
2. data sets having sites that were active during varying time periods, that included a minimum list of constituents based on the water-quality standards that apply

to a particular site and often were collected from an inappropriate sample medium (e.g. analysis of hydrophobic compounds in samples of open water), and generally do not include flow;
3. some computerized data sets, but in varying formats and languages (much of the data were "available" in paper files that required data entry);
4. incomplete data sets (empty fields) containing errors that caused all data to be of questionable quality;
5. inadequately documented sample-collection techniques and laboratory analytical techniques that made it difficult to determine if the techniques used were the same or similar; and
6. few ancillary data sets that contained current information at the appropriate spatial or temporal scale.

Because the existing data were not adequate to meet the goals of the NAWQA program, the study-unit teams designed surface and ground water quality monitoring networks within large river basins that, when combined with existing programs, will provide the information necessary to meet the program goals.

3.1.3. *Intensive Data Collection*

The third phase for each study unit was implementation of the 3-year intensive data-collection component of the study design. Each study unit design included the standardized approaches and data collection, quality assurance, and analysis techniques agreed upon during the planning and design phase [11]. Great care was taken to consistently use those approaches and techniques so that information can be compiled at various scales. For the NAWQA Program to succeed at the regional and national levels there must be a high level of assurance that data were collected and analyzed in a consistent manner to minimize differences due to data collection and analysis procedures. A chemical concentration measured in a water sample from one study unit must be directly comparable with a chemical concentration from any other study unit.

Sample sites were chosen based on a study-unit stratification process whereby basins were subdivided by major features such as physiographic provinces or major aquifer systems that may influence water quality, and further subdivided by predominant land uses. Because of limited resources (funds, personnel, and time) a small number of sampling locations needed to be selected from a large list of potential sites that met the stratification criteria within each basin. Care was taken during the selection of sample sites to assure that data collected from each site helped to address questions and issues at local, regional, and national scales. Also, sample sites needed to be typical of the larger, surrounding area so that information gained through intensive monitoring could be transferred to adjacent locations that were monitored less frequently, and ultimately provide a description of water quality for the study unit.

Surface-water Monitoring Networks. A surface-water monitoring network was developed based on a nested design that included 50-100 sample sites spatially distributed throughout each study unit. Sample sites were chosen to represent various land uses, upstream and downstream of known point- and nonpoint-source inputs, and various scales ranging from small subbasins to large tributaries, to main-stem rivers.

Sample frequency ranged from infrequent sampling of the complete network of sites to determine occurrence and distribution of selected chemicals and compounds in water, sediments, and biota; to a smaller network of fixed monitoring sites that were sampled monthly and during extreme high and low flows; and finally to a subset of the fixed monitoring sites that were sampled weekly during periods of intensive application of nutrients and pesticides. In general, because of analytical costs, samples collected from the larger network were analyzed for a shorter list of constituents than samples collected from the few intensive monitoring sites. The following provides a brief description of types of monitoring sites, their intended purpose, sample frequency, and a list of constituents.

Integrator Sites: A small number (2-4) of fixed monitoring sites were located at continuous-stream-flow gaging stations on main-stem rivers that generally have large drainage areas and mixed land uses. These sites serve as *integrators* of upstream water quality. Water samples are collected monthly and an attempt is made to collect samples during extreme high and low flows [12]. Samples are analyzed for concentrations of nutrients, major ions, suspended sediment, and organic carbon. On-site measurements of flow and field parameters (pH, dissolved oxygen, specific conductance, water temperature, and alkalinity) are made during each site visit. Data from these sites, which often supplement existing monitoring programs supported by other agencies, form the core of the national network and will provide the data necessary to analyze for long-term trends.

Indicator Sites: A second group of 6-10 fixed monitoring sites were located at gaged sites (if a site did not have a gage, one was constructed so that a continuous-stream-flow record was available) in small watersheds (50-270 square kilometers [km^2]) and serve as indicators of the effects of the predominant land use within that watershed. The same sample frequency and constituent list was employed at indicator sites as at the integrator sites. In addition, samples of the fish [13], macroinvertebrate [14 and 15], and benthic algal communities [16] were collected annually and a series of physical measurements were made to describe terrestrial and in-stream habitat [17]. Data from these sites form the core of the national network that will provide information on the effects of land use on water quality at various scales and in different hydrologic settings throughout the United States.

Intensive Sites: A subset of indicator sites (1-4 sites) was selected for *intensive* sampling during periods of heavy application of nutrients and pesticides, primarily in agricultural and urbanized watersheds. Depending on the study unit, samples were collected weekly for one month to several months. Samples were analyzed for concentrations of nutrients, major ions, suspended sediment, organic carbon, and pesticides. On-site measurements of streamflow and field parameters were made during each site visit. These sites form the core of the national network that will provide estimates of flux from different land uses, and insights on the seasonal occurrence and magnitude of sampled constituents that are vital for development of more efficient and effective sampling designs.

Because in some study units the core of fixed monitoring sites represented only a small part of the total land area within the river basin there was concern about the level of confidence in the data, and that without a measure of variability, the information from these sites may have limited transferability to other sites, and to the study unit as a whole. Therefore, additional sites were selected for comparison with the indicator sites to help define variability between sites that had been chosen to represent a predominant land use. Generally, 2-4 comparison sites were paired with each of the indicator sites. Comparison sites met the same site-selection criteria as the indicator sites, but were sampled only 1-2 times per year. Samples were analyzed for the same list of constituents as the indicator sites, but with fewer biological analyses. In addition, 2-3 comparison sites were selected within each indicator basin to help define within-basin variability. Data from these sites are important at the study-unit scale because they provide a level of confidence about transferability of information to the larger scale; however, they are not part of a regional or national network.

Synoptic Sites: Additional sites were selected to provide a more complete spatial coverage of the river basin. These sites represented large tributaries that did not meet the criteria listed above, but represented large areas of mixed land use and provided substantial inputs to mainstem rivers. Sites on main stem rivers that were not selected to be fixed monitoring sites also were a part of this network and represented locations upstream and downstream of major inputs or major influences such as lakes and reservoirs. This group of sites, when combined with all sites described above, became the study unit's synoptic network. All sites in the network were sampled 1-2 times per year during a short period of time (2-3 weeks) by a large number of field crews. Samples were analyzed for the core list of chemical parameters and field measurements (some study units included pesticide analyses) in order to provide a spatial "snapshot" of the water quality within the basin at that time. Because all other monitoring sites were nested within this synoptic network the data provided each study-unit team a means of putting individual sites in perspective within the basin, and provided some level of confidence about the transferability of information from an intensive site to a larger land area that the site was chosen to represent.

Bed Sediment and Tissue Surveys: Each study-unit team selected 20-40 sites from the synoptic network for a survey of metals and hydrophobic organic compounds that may have accumulated in bed sediments and bioaccumulated in tissues. Fine-grained, surficial sediments representing recent deposition were collected; a subsample of the <63μ fraction was analyzed for metals and a subsample of the <2mm fraction was analyzed for organochlorine compounds, semivolatile organic compounds, and organic carbon [18]. Whole-body tissue samples were analyzed for metals and organochlorine compounds [19]. The tissue analyses were intended to provide information on occurrence and spatial distribution of chemicals and compounds that could be used to identify potential problem areas and help to develop the study design, and were not intended to be used for comparison with health-related standards. Target species were prioritized in an attempt to provide standardization nationally. The asiatic clam Corbicula fluminea was the first choice. If Corbicula fluminea was not present, or in low abundance, benthic-feeding fish such as carp, catfish, or suckers were the second choice, followed by species of macroinvertebrates such as caddis flies.

Ground-water Monitoring Network: Ground-water monitoring networks were designed using three approaches that provided information at three scales: 1) a survey of wells and springs distributed throughout the study unit or a major aquifer system; 2) a survey of wells and springs representing the effects of specific land uses on the surficial aquifer system(s); and 3) surveys of transects of wells, springs, and seeps representing localized, small-scale flow systems [20]. The objectives of these surveys, and the emphasis placed on each approach, varied depending on the hydrogeology of the study unit. Generally, sample sites that were part of large-scale surveys were sampled once to determine occurrence and distribution of a large suite of chemicals and compounds [21]. Sample sites that were part of small-scale flow-system studies were sampled more frequently for a smaller suite of chemicals and compounds to provide understanding of processes that influence their fate and transport from specific land uses.

Study Unit Survey: As a part of the retrospective analysis of existing data, the locations and physical descriptions of existing wells were incorporated into the study unit data base. That list was then sorted to provide a subset of 40-60 wells that met specific criteria such as: well constructed to avoid localized contamination, open interval represented the aquifer of interest, well could be sampled using field techniques required for quality assurance, etc. The intent of the survey was to describe water quality of a shallow aquifer or aquifer system that may be used for water supply and may be susceptible to contamination. Generally, the available wells were observation wells that had been installed during previous studies or domestic wells. In areas where few wells were available, springs were used to supplement the monitoring network. In study units having a large number of wells and springs, a standardized computer program developed by Scott [22] was used to randomly select spatially distributed wells for sampling to assure statistical validity of the data.

Land-use Surveys: Land-use surveys were designed to determine the effects of a specific land uses on the surficial aquifer system(s). Often the surficial aquifers do not provide water supplies but their water quality is of interest to determine any effects of overlying land uses, and to provide an early warning of potential contamination of underlying aquifer systems. Generally, the target land uses were urban and agricultural. The standardized computer program by Scott [22] was used to randomly select 20-40 sample locations. If existing wells and springs were in close proximity to the selected locations, and if they met the survey criteria, they were used as sample sites. Otherwise, shallow wells were installed. The preferred approach was to install wells to ensure consistency in well construction and materials. In general, this was feasible and cost effective in agricultural areas. However, in urban areas it was difficult to obtain permission to install monitoring wells, so sampling sites were limited to existing wells and springs that met the established criteria.

Flow-system Studies: Ground-water flow-system studies were designed with the intent of tracking specific chemicals or compounds applied at land surface, through the shallow ground-water system to an area of surface-water discharge. Studies generally focused on transport of nutrients and pesticides applied in agricultural and urban settings, and represented varying hydrogeologic and geographic settings. Unlike the approaches described above, these studies characterize local flow systems that are subject to possible ground-water contamination from land-use practices. A series of nested wells were installed along a transect believed to parallel the ground-water flow path from the source of contamination to the area of discharge to surface water. Springs, seeps, and drains within the study area often were used as sample points to supplement information from the wells and to gain a better understanding of the flow system. Samples were collected at varying frequencies, but generally collected to represent seasonal conditions.

Ancillary Data: Each study-unit team acquired ancillary data sets consisting of information that would help to describe and explain the water quality within the study unit. This information included, but was not limited to: soils, geology, population, land use and land cover, climatic data, and agricultural statistics such as crop types and acreages, fertilizer and pesticide usage, and numbers and types of animals produced in a given area. Much of these data were included in Geographic Information Systems (GIS) to assist with numeric and graphical evaluation.

Special Studies: Each study-unit team had the flexibility to conduct special studies that addressed issues critical to the river basin or major aquifer system, provided the core monitoring networks produced information necessary for regional and national syntheses. Several study-unit teams expanded their study designs by adding sample locations, adding or modifying approaches, or addressing water-quality topics beyond the national focus.

3.1.4. Analysis, Interpretation, and Report Production

The fourth phase of the NAWQA Program is data analysis, interpretation, and production of technical reports and other products such as briefing papers and educational materials. The 20 study-unit teams that began work in 1991 currently (1996) are in the fourth phase of the study. Each study-unit team is preparing topical reports that describe results and conclusions, and are archiving all information for future reference. Teams of scientists at the regional and national level are compiling data and interpretive information from the 20 study units, combining it with data and information from other sources, and will conduct regional and national syntheses that address issues related to nutrients, pesticides, volatile organic compounds (VOCs), and aquatic biota.

Currently (1996) teams of scientists from the first group of 20 study units are collecting samples at a subset of monitoring sites as part of a low-intensity phase of study that is designed to provide a link between the high-intensity phases that will occur in 10-year cycles, and provide a better understanding of long-term trends that may occur. The second group of 20 study-unit teams began the planning phase in 1994, the retrospective analyses of existing data in 1995, and the 3-year intensive data-collection phase in 1996. The third and final group of 20 study units are scheduled to begin the planning phase in 1997, the retrospective analysis phase in 1998, and the 3-year intensive data-collection phase in 1999. This staggered schedule allows the centralized laboratory to receive a consistent sample load, and allows the NAWQA Program to utilize funds and human resources efficiently and effectively. By the time the third group has completed the intensive data-collection phase (2001) the teams from the first group of study units will have evaluated the initial study designs, modified them to accommodate new or evolving issues, and will be prepared to begin a second cycle of intensive data collection. The NAWQA Program is intended to continue to provide water-resource information at the local, regional, and national level indefinitely.

Some of the data collected during the first cycle of intensive data collection have been summarized and are described in reports that have been produced by teams at the study-unit and national level. Other data are being analyzed, interpreted, and documented in topical reports that will be produced during 1996-97. A list of reports published by the NAWQA Program, along with other information, is available at the USGS home page on the World Wide Web at http://www.usgs.gov.

4. Summary

In 1986 the US Congress asked if the Nation's water-quality was improving as a result of the expenditure of an estimated $541 billion for water-pollution abatement in the United States during the period 1972-86. For a number of reasons, this important question could not be answered satisfactorily using existing water-quality information. Therefore, the US Geological Survey was provided funds to design and implement a National Water-Quality Assessment Program to address questions related to status and long-term trends in surface- and ground-water quality at national, regional, and local scales. Information from the NAWQA Program will help support the development and evaluation of management, regulatory, and monitoring decisions by other Federal, State, and local agencies to protect, use, and enhance water resources.

Some of the experience gained by the USGS during the development of the NAWQA Program may be useful when developing an international data-sharing programme for the effective management of Danube River basin resources. It is important for participating nations to agree upon ways to share data, but if those data were collected as part of compliance-monitoring programs, similar to those in the United States, it cannot be assumed that those data alone will provide the information needed to assess the quality of the Danube River, or design effective management programmes. In addition to agreeing to share data, it is equally important to agree upon mutual goals and a valid design for the assessment programme. It is important to start by identifying questions and issues related to the water quality of the Danube River basin. This should be an iterative process that is open to consensus building.

The next step in designing an assessment programme is prioritization of the questions and issues, realizing that an assessment programme may not be able to address all questions and issues, and also realizing that priorities may vary throughout a river basin as large as the Danube. Based on the priority questions and issues to be addressed, a list of data and information needs can be identified and prioritized. This would include locations of key sample sites, constituent lists, sample-collection and laboratory-analytical techniques, sample frequencies, types of ancillary data, and data-reporting criteria. The next step is to identify and locate sources of existing water-quality, water-quantity, and ancillary data; consolidate existing data; process, summarize, and evaluate the data; and share results. This information could then be used to determine which of the priority questions could be answered using existing data, which could not be answered, and what data are necessary to fill information gaps. An assessment programme could then be designed to fill the gaps, taking care to standardize approaches, techniques, and data bases, and to use new data in conjunction with existing data to address the priority questions. Standardization is critical throughout the programme so that the combined efforts of all participants will provide the information needed to answer the questions and, thereby, attain the agreed upon goals.

5. References

1. Intergovernmental Task Force on Monitoring Water Quality (ITFM) (1992) Ambient Water-Quality Monitoring in the United States - *First Year Review, Evaluation, and Recommendations: A Report to the Office of Management and Budget*, Washington, D.C.

Camp Dresser & McKee, Inc. (1986) *The Safe Drinking Water Act as Amended by the Safe Drinking Water Act Amendments of 1986*, Camp Dresser & McKee, Inc., Boston, Mass.National Academy of Sciences and National Academy of Engineering (NAS/NAE) (1973) *Water Quality Criteria*, 1972, US Environmental Protection Agency, EPA R3-73-033.

4. Nowell, L.H., and Resek, E.A. (1994) National Standards and Guidelines for Pesticides in Water, Sediment, and Aquatic Organisms - Application to Water-Quality Assessments: Reviews *in Environmental Contamination and Toxicology*, v. 140, Springer-Verlag, New York.
5. US Environmental Protection Agency (1993) Protection of Environment, revised July 1, 1993 - *US Code of Federal Regulations*, Title **40**, Parts 100 to 149, US Government Printing Office, Washington, D.C.
6. US Environmental Protection Agency (1994) *National Primary Drinking Water Standards, US Environmental Protection Agency*, EPA 810-F-94-001.
7. US Environmental Protection Agency (1995) *Drinking Water Regulations and Health Advisories*, US Environmental Protection Agency.
8. Powell, Mary (1995) Building a National Water-Quality Monitoring Program, *Environmental Science & Technology*, v. **29**, no. **10**, 458-463.
9. Hirsch, R.M., Alley, W.M., and Wilber, W.G. (1988) *Concepts for a National Water-Quality Assessment Program*, US Geological Survey, Circular 1021.
10. Leahy, P.P., Rosenshein, J.S., and Knopman, D.S. (1990) *Implementation Plan for the National Water-Quality Assessment Program*, US Geological Survey, Open-File Report 90-174.

11. Gilliom, R.J., Alley, W.M., and Gurtz, M.E. (1995) *Design of the National Water-Quality Assessment Program - Occurrence and Distribution of Water-Quality Conditions*, US Geological Survey, Circular 1112.
12. Shelton, L.R. (1994) *Field Guide for Collecting and Processing Stream-Water Samples for the National Water-Quality Assessment Program*, US Geological Survey, Open-File Report 94-455.
13. Meador, M.R., Cuffney, T.F., and Gurtz, M.E. (1993) *Methods for Sampling Fish Communities as Part of the National Water-Quality Assessment Program*, US Geological Survey, Open-File Report 93-104.
14. Cuffney, T.F., Gurtz, M.E., and Meador, M.R. (1993) *Methods for Collecting Benthic Invertebrate Samples as Part of the National Water-Quality Assessment Program*, US Geological Survey, Open-File Report 93-406.
15. Cuffney, T.F., Gurtz, M.E., and Meador, M.R. (1993) *Guidelines for the Processing and Quality Assurance of Benthic Invertebrate Samples Collected as Part of the National Water-Quality Assessment Program*, US Geological Survey, Open-File Report 93-407.
16. Porter, S.G., Cuffney, T.F., Gurtz, M.E., and Meador, M.R. (1993) *Methods for Collecting Algal Samples as Part of the National Water-Quality Assessment Program*, US Geological Survey, Open-File Report 93-409.
17. Meador, M.R., Hupp, C.R., Cuffney, T.F., and Gurtz, M.E. (1993) *Methods for Characterizing Stream Habitat as Part of the National Water-Quality Assessment Program*, US Geological Survey, Open-File Report 93-408.
18. Shelton, L.R., and Capel, P.D. (1994) *Guidelines for Collecting and Processing Samples of Stream Bed Sediment for Analysis of Trace Elements and Organic Contaminants for the National Water-Quality Assessment Program*, US Geological Survey, Open-File Report 94-458.
19. Crawford, J.K., and Luoma, S.N. (1993) *Guidelines for Studies of Contaminants in Biological Tissues for the National Water-Quality Assessment Program*, US Geological Survey, Open-File Report 92-494.
20. Lapham, W.W., Wilde, F.D., and Koterba, M.T. (1995) *Ground-Water Data-Collection Protocols and Procedures for the National Water-Quality Assessment Program - Selection, Installation, and Documentation of Wells, and Collection of Related Data*, US Geological Survey, Open-File Report 95-398.
21. Koterba, M.T., Wilde, F.D., and Lapham, W.W. (1995) *Ground-Water Data-Collection Protocols and Procedures for the National Water-Quality Assessment Program - Collection and Documentation of Water-Quality Samples and Related Data*, US Geological Survey, Open-File Report 95-399.
22. Scott, J.C. (1990) *Computerized Stratified Random Site-Selection Approaches for Design of a Ground-Water-Quality Sampling Network*, US Geological Survey, Water-Resources Investigations Report 90-4101.

ENVIRONMENTAL POLICIES IN TRANSITION:
A NATO Partnership Country Symposium

IRENE LYONS MURPHY
Colorado State University
2005 37th Street NW, Washington, DC, USA

KEYWORDS / ABSTRACT: environmental legislation / Danube River Basin / Eastern Europe / water quality monitoring / water quality data-sharing

Representatives of the Danube countries in transition presented information about their respective water quality monitoring and data-sharing policies and programmes. Some reported general satisfaction with the support programmes were providing government decision makers. Others saw the need to improve the quality of monitoring as well as the use of electronic databases and network to facilitate improved environmental management. The presentations by the eastern Danube countries generated a discussion about current monitoring databases.

1. Introduction

Three of the NATO countries reporting on their water quality monitoring and data-sharing policies in the previous chapter have harmonised their national environmental legislation under European Union Council directives. Germany, the Netherlands, the United Kingdom as well as the United States can be said to have adopted stringent environmental standards and helped to implement them with the use of state-of-the-art monitoring and data reporting technologies. All of them are party to international agreements with neighbouring countries which mandate some form of transboundary monitoring and data-sharing. And all have responded to steadily growing demand for the publication of water quality and other environmental data from organised constituencies and the general public.

The European Union members of the Danube Basin countries, Austria and Germany, have enacted environmental legislation which, in general, harmonises with other western European countries. The remaining Danube countries, save for Serbia and Bosnia, have embarked on a transition period which includes the development of environmental policies in harmony with those of the west. The Czech Republic and the Slovak Republic, Slovenia, Hungary, Bulgaria and Romania, as well as Moldova and Ukraine are NATO partnership countries. The six East European countries have applied for membership in the EU. Croatia also participates in Danube co-operative programs and is modifying its legislation to make it compatible with western norms.

I. Lyons Murphy (ed.), Protecting Danube River Basin Resources, 105–111.

Incentives include a desire for increased trade with and investments from western sources, and to follow through in other ways with commitments accompanying applications to join the European Union. A number of steps have been taken by the Environmental Programme for the Danube River Basin (EPDRB), [see Rodda, 113 and Lack, 127] to support revised policies during the transition period. These include the enactment and implementation of legislation with a range of objectives, including compatible water quality standards and enforcement and the sharing of water quality monitoring data. The Monitoring, Laboratory and Information Management Study Group project has completed a number of steps in this direction. Objectives include the strengthening of existing environmental monitoring networks, improvement of the comparability of sampling techniques and laboratory analyses, and development of information management systems to serve decision makers at the local and international level.

As a follow-up to the previous chapter, which included details of four NATO country water quality monitoring policies, representatives of ministries of environment or of research institutes familiar with legislation and data-sharing policies from Eastern bloc countries were asked to respond to a brief questionnaire about current policies and their effectiveness. They were also invited to bring brief statements about them to the meeting and to be prepared to discuss them with the objective of determining their general effectiveness and further need for change.

Participating in the discussion were: **Ivan Milushev**, Ministry of Environment, Bulgaria; **Željko Ostojič**, State Water Directorate, Zagreb, Croatia; **Jan Kubat**, Czech Hydrometeorological Institute, Prague, Czech Republic; **Eszter Poroszlai**, Ministry of Environment Protection and Regional Policy, Budapest, Hungary; **Corneliu Negelescu**, Research and Engineering Institute for Environment, Bucharest, Romania; **Peter Hlavatý**, Ministry of Environment, Slovak Republic; **Meta Gorišek**, Ministry of Environment and Physical Planning, Ljubljana, Slovenia; **Volodimir Bilokon**, Ministry for Environmental Protection and Nuclear Safety, Kiev, Ukraine. **Ilie Dediu**, National Institute of Ecology, Kishinev, Moldova, had responded to the questionnaire but was unable to come to the workshop.

2. Are Water Quality Data-sharing Policies Adequate?

All countries represented at the panel had replied to the questionnaire. The question was: Does water quality data now collected in your country and from other countries satisfactorily support ministry short and long-term decisions and those of other government agencies? Questionnaire results come first and are followed by the respective statements of the East European country representatives.

BULGARIA

Ilya Natchkov, Bulgarian Ministry of Environment, National Centre for Environment and Sustainable Development, represented by Ivan Milushev, had responded that Bulgarian policies which have been in effect for decades have developed extensive water quality monitoring databases. He believes that there could be improvements in database management; a computerised decision support system is needed and that communications tools for data sharing and transmission could be improved. Mr. Milushev provided a brief history of legislation and the administrative structure for environmental management. Future plans include the acquisition of a telephone line to transmit data from the Regional Environmental Inspectorate to the National Centre and the Ministry of Environment. They are also developing computer networks in the Centre to connect with recently distributed computers.

CROATIA

Željko Ostojič, State Water Directorate, Zagreb, Croatia, stated that present data collection did influence the adoption of water management programs regarding water protection from pollution, changes in legislation regarding limitation of waste water discharging, and conditions for issuing water management licenses. He provided the panel with a description of the recently enacted Water Act (December 27, 1995). The legislation supports improved reporting on water management conditions. There are inventories of surface and ground water conditions, the construction of water management projects and water supply facilities. Water protection inventories record data on pollution sources, waste water drainage systems and other facilities. Data from these and other inventories are available to the public. At present Croatia and Hungary have an agreement to share information on water balance, and water use, as well as protection from pollution.

Mr. Ostojič concluded his statement by saying that in the Croatia part of the Danube much information was gathered that is not yet automatically processed because of lack of funds, necessary legislation, and the lack of qualified personnel to work on computers. Mr. Ostojič pointed out that as a young state, the Republic of Croatia is beginning only now to establish contacts with neighbouring countries so as to exchange relevant water management information.

CZECH REPUBLIC

Jan Kubat, Czech Hydrometeorological Institute, Prague, Czech Republic, suggested that present policies did support decisions as well as they could. The Ministry of Environment finances the water quality monitoring and receives monitoring results. Water quality data are used for determining the state water management balance, water management planning, assessment of water part of environment and state administration generally. Large rivers are in the hands of the Ministry of Environment and its Water River Enterprises agency and smaller rivers under the purview of the Ministry of Agriculture and Forestry. Approximately 280 stations comprise the state monitoring network for water quality under the Czech Hydrometeorological Institute. Presently only two sites are part of the international monitoring network (for the Morava and Dyje Rivers. Mr. Kubat thinks it may be unfortunate that market principles are introduced into monitoring. The state budget does not fully cover the costs so that data exchange between institutes may carry a charge. Rules for charging are unclear; he suggests the need for clarifying them.

HUNGARY

Eszter Poroszlai, Ministry for Environment and Regional Policy, Budapest, saw special problems with respect to information about water quality from other countries particularly since the Helsinki and Danube Conventions specified that such exchange should take place. The Hungarian system, she believes, is rich in data but poor in information. The standardisation of data to satisfy international needs was begun two years ago. Ninety-five per cent of Hungary's water comes from outside its borders and therefore there is need for data exchange but also to resolve some outstanding issues. Hungary needs bilateral agreements to facilitate the improvement of data from the upstream catchment area. Presently data is exchanged only at border sites.

MOLDOVA
Ilie Dediu, National Institute of Ecology, Kishinev, Moldova, suggested that water quality data needed to be more efficiently assembled, the number of parameters measured larger in number and monitored with more efficiency there could be a larger number of monitoring stations, especially on the smaller rivers.

ROMANIA
Corneliu A. L. Negulescu. Research and Engineering Institute for Environment, Bucharest, Romania, amplified statements in the response to the questions about the adequacy of present monitoring and data sharing policies. Such activities are in the hands of the Ministry of Water, Forestry and Environmental Protection and this centralisation has worked well. Water supply is managed by national water authorities within 16 sub-divisions. An annual report of water conditions in Romania in terms of national water quality standards, is provided. Internationally, data is shared with Ukraine, Hungary, and Moldova. Present monitoring is considered to be sufficient for support of the strategic action plan. It is anticipated that EU requirements for such data will be met by the year 2000. Romania is interested in visits by western experts to make investment studies on Danube tributaries.

SLOVAK REPUBLIC
Peter Hlavatý, Ministry of Environment, Bratislava, Slovak Republic stated that the Ministry of Environment has principal responsibility for decisions concerning the environment but that several other ministries also participate in decision making. The reorganisation of the Ministry in 1994 created a new Division of Environmental Economy. The reorganisation allows the Ministry of Environment to more clearly establish priorities and to manage issues more effectively. Analyses of samples are carried out by the River Basin State Enterprises management water quantity and quality in four basins, the Danube, Vah, Hron, and Bodrog and Hornad. The Czech and Slovak republics use the same water quality assessment standards. Generally, Mr. Hlavatý believes that water quality sampling is effective but parts of the current legislation which was enacted in 1975 need changes, particularly with reference to parameters and areas not now included.

SLOVENIA
Meta Gorišek, Ministry of Environment and Physical Planning, Ljubljana, Slovenia, reported that there were too few monitoring stations and that there was inadequate monitoring coverage in some areas of Slovenia. Sometimes there are too much data in some areas and not enough in others. Improvements were taking place as the result of a

new Water Act. These include updating of the monitoring system and adoption of EU water quality and laboratory methods. She reviewed the present monitoring network of inland waters and the plan to integrate monitoring with data on water quality standards. International monitoring takes place on the Austrian, Hungarian, Croatian and Italian transboundary sites. Agreements to monitor and share data are in effect with each of these countries

UKRAINE

Volodimir Bilokon, Ministry for Environmental Protection and Nuclear Safety, Kiev, Ukraine saw a need for improvement in data collection and the processing of a greater degree of information exchange between countries. In recent years legislation has modified environmental policies implemented by the Ministry of Environmental Protection. These include rules for monitoring, implementation of the Helsinki and Danube conventions, and other measures to harmonise Ukrainian environmental management processes with those in Europe and America. The latter in particular will include emission standards which Ukraine has not had. An annual statistical report on water use will also be required. Those discharging pollutants are required to report the quality of water discharges. These and other conditions are annually reported in a state of the environment report. Ukrainian policy is to save the existing monitoring network and improve it especially in view of economic conditions. Ukraine has agreements with Slovakia and Hungary regarding the Tisza River and anticipates making one with Romania later this year.

Participants were assured of the interest of the EPDRB PCU through the presence of Mr. David Rodda who had been invited to the workshop while he was team leader of the PCU. Mr. Rodda said he was still representing the PCU as a programme. Irene L. Murphy, USA, said she would be reporting information on the workshop directly to the new team leader, Teun Botterweg on Friday, 7 June 1996. She also pointed out that Tim Lack, head of the MLIM Sub-Group was a workshop participant.

Aurel Varduca, Romania, and others raised the point that water quality data had to be quality data, properly stored and retrievable. He also said that resulting information needed to be available and shared at all levels. Tim Lack, UK, said that the MLIM goal was to improve the capacity of Danube countries to produce quality data and then transfer it using the DEFF. The exchange of more general data and information is incomplete and the workshop should perhaps focus on both.

David Rodda, UK, said that he looked at it that way. There are many sets of data which have not been used in the EPDRB sets. There are sets which we do not know about concerning, for example, internationally shared data sets. We do not know enough about how these are exchanged internationally. [The goal of several case studies which conclude this volume was to explore this type of "field" data exchange.] Mr. Rodda pointed out that there are many past and current research studies which are not available in a database and that this would be useful.

Hans Liebscher, Germany, suggested that transboundary monitoring on the Elbe was served by the selection of national stations that would also be internationally useful. This had saved money which was a concern in the Danube Basin. Tim Lack, UK, said he did not entirely agree. He said that some national sites might be selected and had been for the Danube. National sites had been selected for national objectives and so it had been necessary to design an international system with its own objectives. It may be necessary since you have dual objectives for the same site to make slight adjustments in accuracy and precision to satisfy the international requirements. But by and large a transnational boundary monitoring network is based entirely on existing stations. After one year the country will be in a position to say whether international requirements are met.

Dr. Lack said that in his work with the European Environmental Agency he had operated as he had on the Danube: each country was asked to provide data from national networks regarding sampling frequency, parameters, etc. A database of databases or a metadatabase was thus formed. A database of legislative regulations and bilateral agreements is also needed. This could produce a catalogue of databases or a meta metadatabase which acts as a routing system. So if you want to know what is the water quality in the Karstic region of Austria, they can tell you, because they know from the database they have from 5 years, from x number of stations, and if that person is interested they can find their way through this information which is publicly available in Austria.

Each topic centre is concerned with, for example, air, land, etc., everyone does it the same way so anyone can have this information which is in a large catalogue. It is really just an electronic library with information about finding out what free information is available, where, etc.

There are so many strands for different schemes that we need to pull them together. Some are on paper, some are electronic on computer. We should try to pull things together to create a catalogue system for the Danube River Basin. We need to compile information. and wouldn't it be nice for that catalogue with a common design, with common architectural rules, to sit beside the western European catalogues. It could be a nice objective to work toward. Could be the first attempt at defining a data source.

Eszter Poroszlai, Hungary said that there are a lot of metadatabases in which you can find which data is where (for Danube Programme, for Europe, for Helsinki, OECD catalogue) All countries give this information to several agencies, but the problem is that for example, BOD isn't measured the same everywhere. It is a question therefore of data compatibility.

Tim Lack agreed but said it did not nullify the concept. You could put in a warning flag, adding it to the data. We all know about the problems of compatibility but because you can't compare data doesn't mean you shouldn't know where to get data.

For Hans Liebscher this raised a question about the task of the workshop if so much had been done already. David Rodda said that the difficulty was that the collection of data has not been interpreted. He is not sure that the new Danube Convention Secretariat would encourage the completion of a large database. The issue is currently under a lot of discussion because of the ongoing review of programme activities as they compare with Convention requirements so that the secretariat can understand to what degree they can take over work in the program.

The leaders of the PCU task force and the Commission will debate this matter. A July meeting should reach some decisions. The concept of metadatabases seems to be an area of fruitful work to be undertaken. David Rodda said that he had started to make a list of parameters and who covers what. It included inventories. An area not discussed at all is environmental health especially as it relates to drinking water. We have to appreciate economic constraints. We need appropriate mechanisms and must work on a step-by step basis. We haven't been able to get hold of this information from World Health Organisation.

Mr. Leibscher said that the European Community is a member of the Danube Convention and in this regard the committee is participating in decisions of this group. It also provides funding. The commission is giving financing to this but all things must be defined by the commission itself. We can only say that research is needed in this field, etc. but all decisions need to be made by the commission itself.

Atanas Paskalev, Bulgaria, asked Mr. Rodda his opinion of the strategy for financing environmental health issues especially concerning change and development in riparian countries? Is there any plan for planned investment?

Rodda replied that it was currently under discussion. There is a Ministerial meeting in the autumn but it will not include all the ministries. Another question is whether or not the financing institutions will go along with a proposal to fund. The World Bank, EBRD, respond cautiously but the idea is still on the table. Atanas Paskalev replied that all matters end with financial situations.

Péter Bakonyi said that if we want a program funded then we need a feasible proposal, something useful for financing institutions.

The panel was not intended to provide a detailed report on monitoring and data-sharing activities in each country because of time constraints and also because the reports of the EPDRB MLIM Sub-Group have covered many aspects of the topic in its reports. The intent rather was to support the adoption of recommendations during the final phase of the workshop. It raised significant issues which are be referenced in the report on the final session.

DEVELOPMENT OF INFORMATION ACTIVITIES IN SUPPORT OF THE ENVIRONMENTAL PROGRAMME FOR THE DANUBE RIVER BASIN

D.W. RODDA
Adviser, Danube Programme Co-ordination Unit
1, The Chestnuts,
Walton On Thames,
Surrey KT12 1EE
UNITED KINGDOM

KEY WORDS / ABSTRACT: Danube River Basin / water quality databases / environmental programmed for the Danube River Basin / environmental information strategy

The scope and work of the Environmental Programme for the Danube River Basin (Programme) and the provisions of Danube-related Conventions, Agreements, or Declarations which give rise to exchange of information are described. Programme achievements are highlighted as well as future activities. Data and information strategy are examined and a comprehensive table of the elements of such a strategy describe. Public information must be systematically addressed. Conclusions are reached about the assignment of responsibility to the Danube countries for data and information dissemination, the advantages of network links between systems rather than designing and implementing new system(s), and the cost-benefits of such activity.

1. Introduction

The Danube River basin is some 817,000 sq. kms in area and from Donaueschingen in Baden Württemburg in Gemany to the Black Sea the distance along the main river is 2857 kms. A current estimate of the total population is near to 85 million. As many as 9 riparian countries receive aid from international agencies for environmental improvements. Certain countries (3 or 4) of the Former Republic of Yugoslavia might be added in the future. The 9 do not include Switzerland, Poland, Italy and Albania, each of whose territorial area within the catchment is less than 2% of the total.

New legislation leading to improvements in environmental measures has been introduced in most of the countries since 1991. National Environmental Strategies have been approved. These form the basis for the National Environmental Action Plans which countries agreed to produce following the Conference of Environment Ministers of Central and Eastern Europe held at Lucerne in April 1993. Further direction to the

I. Lyons Murphy (ed.), Protecting Danube River Basin Resources, 113–125.

measures required to improve the environment was given at the Conference of Environment Ministers of Central and Eastern Europe held at Sofia in October 1995.

Within the framework of activity started in 1991, the Danube countries and the International Community agreed on a special short-term Programme of assistance to support the countries own co-operation to improve environmental management. This Programme known as the Environmental Programme for the Danube River Basin (Programme) started in 1992.

At about the same time the countries proceeded to negotiate two Conventions or Agreements: one to address the principles of Water Management, the other the Protection of Ecosystems. The Water Management Convention, known as the Danube River Protection Convention, follows the principles embodied in the Convention on the Protection and Sustainable Use of Transboundary Waters and International Lakes signed in Helsinki in March 1992. The Danube countries signed the Danube River Protection Convention in Sofia in June 1994. To date 4 countries have ratified the Convention: 9 signatories are required before it enters into force. Negotiations on a Danube basin legal instrument to address the protection of Danube ecosystems continue.

2. Information Provisions in Danube-relevant Conventions or Agreements

2.1 SOFIA DECLARATION

The Declaration, drawn up at the 1995 Sofia Environment Conference - the third in a series which started with the Dobris Castle Conference in 1991 - reaffirmed the overall commitment by Central and Eastern European countries to cooperate in the field of environment protection. This commitment was linked to economic growth. Some of the main points covered are:

- Continuation of progress to implement the Environmental Action Programme for Central and Eastern Europe was endorsed.
- A major report on environmental financing was welcomed and countries were encouraged to adopt the approaches outlined in order to implement the "Polluter Pays Principle".
- The significant role that the business, industry and environment sector can play in securing sustainable development was recognised.
- The uniqueness of biological and landscape diversity was underlined and the Pan-European Biological and Landscape Diversity Strategy was endorsed.
- The need for protection from nuclear risks was emphasised.
- The value of the "Dobris Assessment" published by the European Environment Agency was endorsed.
- This publication and the database it presents underlines the importance of information exchange as an enabling process for improving environmental management. The European Environment Agency was asked to build on this work using a Pan - European network (to be developed) for data collection, analysis and dissemination.

- The need for improved public participation and the important role of regional environment centres was endorsed.

2.2 ENVIRONMENTAL PROGRAMME FOR THE DANUBE RIVER BASIN

The objective of the Programme is short-term. Its aim is to establish an operational basis for strategic and integrated management of the Danube River Basin environment. The work carried out is steered by a Task Force specially set up for the purpose.

A main operational principle of the Programme is to encourage good co-operation in order to develop a consensus on the priority work which has to be carried out and to promote information exchange. A Programme Co-ordination Unit (PCU) established to run the Programme on a day-to-day basis, was, *inter alia*, given a responsibility to facilitate information sharing. Reference is made to the importance of data management with the objective of developing appropriate tools and methodologies.

2.3 CONVENTION ON THE PROTECTION AND USE OF TRANSBOUNDARY WATERS AND INTERNATIONAL LAKES

This Convention was negotiated as a component of the work programme of the UN Economic Commission for Europe's (UN ECE) Senior Advisers on Environmental and Water Problems now renamed the Committee on Environment Policy. The Convention was signed in Helsinki in March 1992 and is regarded as the "father" of the Danube River Protection Convention.

The objectives of the Convention are to require the signatories to take all measures to:

- Prevent, control and reduce pollution of waters causing or likely to cause. adverse transboundary impact,
- Ensure that transboundary waters are used with the aim of achieving ecologically sound and rational water management,
- Conserve water resources and protect the environment,
- Ensure that transboundary waters are used in an equitable way,
- Conserve and, where necessary, restore ecosystems.

The Convention contains an Article (No. 6) on "Exchange of Information". It states that "Parties shall provide for the widest exchange of information, as early as possible, on issues covered by the provisions of the Convention." In a later Article (No. 8) entitled "Protection of Information", there is a provision which states "The provisions of this Convention shall not affect the rights or obligations of Parties.........to protect information related to industrial and commercial secrecy, including intellectual property, or national security."

This Convention has not yet entered into force, but 27 countries have signed and 14 have ratified.

2.4 CONVENTION ON THE PROTECTION AND SUSTAINABLE USE OF THE DANUBE RIVER

The Convention (the short title is Danube River Protection Convention) signed at Sofia in June 1994, aims at achieving equitable and sustainable water management. Its particular objectives are:

- to "control hazards originating from accidents involving substances hazardous to water, floods and ice-hazards",
- the "conservation, improvement and the rational use of surface and groundwaters in the catchment area",
- to "contribute to reducing the pollution loads of the Black Sea from sources in the catchment area."

Provisions for the exchange of information are included in Article 4 in which "exchange of information to establish a legal base is entered as a form of co-operation."

Article 12 deals quite specifically with "Exchange of Information". The provision states: "Contracting Parties shall exchange reasonably available data" on a range of items e g riverine environment, experience with best available techniques, research and development, emission and monitoring data, measures planned, regulations for wastes discharges, and accidents. Provision is made for Contracting Parties to make reasonable charges for exchanging data and information. The rights of Contracting Parties in accordance with their domestic laws are protected however. This aspect is specifically covered by Article 13 - "Protection of Information supplied".

Other relevant Articles which imply an exchange of information are:

- No. 9 which provides, in respect of monitoring, for Contracting Parties to "develop concerted or joint monitoring systems applying stationary or mobile measurement devices, communication and data processing facilities." This is followed with wording requiring the elaboration and implementation of "joint programmes for monitoring".
- No. 10 which sets out a list of "Obligations for Reporting". It is to be presumed that the reports involved will be available to the Contracting Parties as members of the International Commission and be published.
- No. 11 which refers to a "prior exchange of information" in order to prepare for consultations on the planned activities likely to cause transboundary impact.
- No. 14, entitled "Information to the Public", which provides for the competent authorities of the Contracting Parties to make information available concerning the state or quality of the riverine environment.

To date no interpretation of these Articles has been discussed by the interim International Commission during its meetings. Thus open discussion on possible interpretations may not be conclusive. There are also other Conventions which apply in the Danube River basin, e g the Convention on Wetlands of International Importance especially as wildfowl habitats (Known as the Ramsar Convention) but their scope is not discussed in this paper.

3. Programme Achievements

During the period to the end of 1995 the following results or progress have been achieved:

- preparations are almost complete to commission a purpose-designed Accident Emergency Warning System to provide for warnings of sudden pollution events, building on experience in the Rhine catchment,
- capacity-building to improve environmental management and operational activities at the National and local levels has covered most environmental aspects. For example there is, under way, a major effort to improve municipal wastewater treatment management and there has been a successful development of a Danube basin-wide NGO network as a firm base for improved public participation. All this is on-going, and, to be sustainable, success in the long-term will rely on sufficient persons of the right ability to be trained and who can train their colleagues,
- environmental problems and their priorities requiring attention e g point and non-point source pollution, have been comprehensively assessed. Independent assessments were commissioned at the start of the Programme on the state of the environment in the catchment area. These included a series of comprehensive National reviews prepared by each country, a series of diagnostic missions in selected tributary catchments covering nearly 50% of the catchment area. a major catchment-wide assessment of the impact of non-point source pollution, and recommendations on the design of an inventory for biological resources,
- each country's capability for monitoring, laboratory and information management has been assessed. A monitoring strategy appropriate to defining transboundary impact has been defined and is being implemented. The end result should be a set of high quality water environment data which builds on and complements the Bucharest Declaration (1985) data,
- a Strategic Action Plan based on the information made available from the Programme activities so far was completed in 1994. Preparation of the Plan was an intensive process throughout the year. The Plan was endorsed in Bucharest in the December of that year by Danube Environment Ministers and the European Union Commissioner responsible for the Environment. The goals of the Plan are:

  a) to reduce the negative impacts in the Danube river basin and on riverine ecosystems and the Black Sea,
  b) to maintain and improve the availability and quality of water on the Danube River Basin,
  c) to establish the control of hazards from accidental spills,
  – to develop regional water management co-operation.

- A major Applied Research programme among co-operating Danube Research Institutes has been commissioned costing 4.5 million ECUs. It comprises 13 projects lasting between 6 months to 4 years i e to 1999,

- A process for enhancing data management throughout the Danube basin has been established. An information system has been designed and named DANIS, which contains textual data and information on human resources in the basin such as Institutions active in water environment management and scientific research, experts and their involvement in the Programme, details of Programme events and other elements of information. A "site" for the database has been launched on the INTERNET.

4. Future Programme Activities

A major effort is under way to define the elements of a Danube Environmental Programme II known as the Strategic Action Implementation Plan (SIP). Once prepared, hopefully by the autumn, and funding is confirmed, all will be ready to commission the work. SIP will contain:

- activities to demonstrate action to implement the Strategic Action Plan by tackling priority problems in selected transboundary tributaries using an integrated management approach;
- continued action to implement a basin-wide monitoring system;
- continued action to enhance the work already completed to commission an Accident Emergency Warning System which will aim at reducing risks that lead to accidents from sudden pollution and the relevant preventative action, risk management, flooding and ice hazards;
- activities to demonstrate appropriate wetlands rehabilitation at selected sites;
- continued work to improve capacity building at the national, municipal, local and NGO levels.

5. A Data and Information Strategy

No management strategy is complete without an information strategy. Both data and information are vital to effective regulation and future planning for the implementation of environmental protection measures. The legal base described above affirms this need.

Initially the Programme marshalled a wide variety of data and information and made assessments on the extent of the concern over the degradation that had taken place and proposed options for their solution. The results are embodied in the Strategic Action Plan and are available in the PCU library. Part of the work to implement the Plan is to develop and apply a series of management tools which can interpret the data and information that will continue to be obtained.

An information strategy is emerging, but it is at an early stage. Table 1 indicates possible components showing some information types of significance.

Activities under the Programme which influence information management and undertaken so far are:

5.1. EXISTING SYSTEMS

First, existing information in the form of documents on national legislation, regulations in force, human resources information and standards set is available on DANIS. Other textual data such as reports on research (fundamental or applied), findings of surveys and learned papers fall in this category. The aim here is to encourage the dissemination of information on the state of the environment and make it available for further analysis and presentation.

In respect of environmental standards, a special database has been established by the European Bank for Reconstruction and Development. The database provides quick access to comprehensive data on environmental standards in 22 countries, Council Directives of the European Union and criteria recommended by the World Health Organisation.

The PCU's library of Danube information in books and reports has also been included in the form of abstracts. Such a database can be used to research and draw conclusions on a wide range of issues including transboundary ones. Another library exists in the Regional Environment Centre, Budapest on aspects relating to Central and Eastern Europe. Other libraries exist in Danube Institutes.

Another type of information, namely "metadata" can be included. Metadata usually describes discrete sets or collections of data. A clear description of individual data items would be included, where the set is kept and in what form, how it can be accessed, whether there any restrictions on access or use such as confidentiality, whether there are any costs to obtain the data and the co-ordinates of the location of the data to facilitate enquiries.

Before the Programme started in earnest, some countries were already responding to international advice on the benefit of information systems. As a result an early project launched a system named INFODANUBE. This is a comprehensive system described as holding and providing ecological data on the Danube environment. It had originally been developed at the Bulgarian Academy of Sciences with funding from the American Fulbright programme. Databases in the system covered, for example, contacts in Ministries and environmental Institutes; water quality standards applied in the basin together with criteria recommended by the European Commission and the UN Economic Commission for Europe; and Bucharest Declaration Monitoring data.

5.2. MEASURABLE DATA

Second there are measured data which require a high degree of quality control for the data to be reliable and therefore useful.

Activities of the countries in the implementation of the Bucharest Declaration (1985) have led to the availability of annual sets of data on Danube main river water quality.

Within the Programme a particular emphasis has been placed on establishing a monitoring strategy for an enhanced routine collection of data on surface water chemical and biological quality. Organisations such as ICWS of the Netherlands and Equipe Cousteau were active and specific data collections are available. Some useful data was obtained which after analysis augments the information available from the work being carried out under the Bucharest Declaration (1985).

It is known that in support of the responsibilities of the Danube Convention(1948) known as the Belgrade Convention, and in support of the Danube as an effective navigable waterway, a significant hydrological database is available at the Vituki Institute, Budapest. Other literature may be available in this context too.

In addition, there are significant sets of measured data available in the Danube countries. These are usually held at the Institutes which collected them. There is no single source of information on these data sets and their availability, but their existence becomes known when specific studies are commissioned e g through the Applied Research Projects.

5.3. FROM DATA TO INFORMATION SYSTEMS

The third aspect is the development and application of management tools to transform available data into information. These often involve the application of mathematical relationships.

The tools allow information to be produced which management requires for decision-making, such as pollution loads moving down a river, or regulation, which involves making decisions on the siting of discharges and the issue of relevant permits. In turn the information supports decisions on the action (type and extent) required to control water quality and the need for additional monitoring. Yet another example is the production of catchment management plans. These facilitate decisions on the equitable use of a river system based on full appropriate information on river use and demands made. A project to demonstrate the value such plans have is currently in progress.

Examples already applied or being applied and known to the Programme are summarised in Table 2.

6. Availability of Information to the Public

The public should have access to information on the environment in which they live. In addition projects to improve the environment will be more successful if public opinion is in favour, especially in cases where domestic financing is required. For that reason the Programme has encouraged the involvement of Danube NGOs and, indeed, has committed funds to provide them with support to mount their own projects and to enhance their networks. A notable example is the setting-up of Information Centres by NGO groups in the Danube catchment area.

Such vital activities thrive on the availability of free information. Thus the PCU facilitates the dissemination of information to the public via the NGO networks and encourages the Danube countries, as do the international financing institutions and donors. Environmental managers require support from the public as it is the surest way to ensure projects benefit the community and that people will pay willingly. Nevertheless agreeing that information is readily available is one thing, making it available is another and public responses may be limited.

7. Conclusions

The above information demonstrates some of the progress being made in the Danube catchment area in respect of information exchange. It will be seen that a full strategy for information exchange has yet to emerge, but progress is being made. The pace at which this proceeds depends very much on the will of the Danube country Governments and donors to make resources available to support such a strategy. In turn the interest of organisations and persons which have responsibility or interest to learn more and investigate environmental situations will demand more and more information.

Experience suggests that to implement a new system throughout the region is, in itself, a major task and should not be undertaken without a careful analysis of the resources required for such items as the appropriate institutions and individuals to be targeted for its implementation and the training and supervision required. To date such an approach does not seem necessary, one of the constraints being the lack of finance. Any funds available from international donors, because it is so precious, must be applied in such a way that the process is sustainable.

Handling information is expensive. The Task Force for the Programme, and the donors of the PCU, will require to be convinced of the benefits of a major activity to strengthen information exchange per se. An example of concern is the procurement and implementation of a uniform data management system based on a suitable GIS. So far funds for a tailor-made Danube system are not forthcoming. In the circumstances it is hoped that the European Commission PHARE support of the extension of CORINE to the countries of Central and Eastern Europe may provide the answer.

8. Acknowledgements

The information used in this paper is drawn from material available from the donor-funded activities of the Environmental Programme for the Danube River Basin. The role of the donors in funding the Programme without whom the work would not have been possible is acknowledged. The opinions expressed in the paper are those of the author and do not represent any views held by Task Force members. Grateful support from colleagues in the Programme Co-ordination Unit for the Programme and colleagues in the Task Force and the participating Danube countries are hereby acknowledged.

9. References

1. Environmental Programme for the Danube River Basin. February 1992. *Work Plan*. Unpublished.
2. Environmental Action Programme for Central and Eastern Europe Abridged version. March 1994. Unpublished

3. Convention on the Protection and Use of Transboundary Watercourses and International Lakes. Signed in Helsinki, March 1992.
4. Convention for the Protection and Sustainable Use of the River Danube. Signed in Sofia, June 1994.
5. Sofia Declaration. October 1995. Endorsed at the Conference of European Environment Ministers held in Sofia.
6. Delft Hydraulics. September 1994. *Accident Emergency Warning System (AEWS), Final Report on Set-Up of System.* Produced under CEC/PHARE Contract for the Task Force expert AEWS Sub-Group.
7. WTV Consortium. February 1994. Monitoring, Laboratory and Information Management (MLIM) for the Danube River Basin. Produced under CEC/PHARE Contract for the Task Force expert MLIM Sub-Group. *Summary Report*. **Volume IV**.
8. Environmental Programme for the Danube River Basin. Strategic Action Plan for the Danube River Basin, 1995-2005. December 1994. *Produced by a special drafting group appointed by the Task Force.*
9. Environmental Programme for the Danube River Basin, Programme Co-ordination Unit. *Programme Annual Reports for 1993, 1994 and 1995.*
10. Environmental Programme for the Danube River Basin. *Final Report on an Information System on Human Resources of the Danube Countries.* January 1995.
11. Institute for Water Quality and Waste Management, Technical University of Vienna. November 1995. *Nutrient Balances for Danube countries and Options for Surface and Ground Water Protection,* Inception Report for an Environmental Programme for the Danube River Basin, Applied Research Project, funded by CEC/PHARE.

Table 1 Candidate elements of an Information Strategy

Textual data	*Collection, storing and retrieving* relevant data.
Library, Regional Events Diary, Meeting attendance, Environmental Standards, Applied Research reports; Human resources data, biodiversity, various inventories e g projects, lists of environmental hot-spots, "metadata".	Operation of a highly flexible networked *common computer database system (DANIS)*, based on CDS/ISIS developed by UNESCO, with well specified data formats for various types of data entry conforming with Internationally acceptable data exchange formats, and user-friendly interrogation routines for information search and retrieval. Provision of *access* via various computer media and INTERNET.
Measured data and statistics	*Collection, storing and retrieving* data and statistical information with a high degree of quality control at all stages.
Water environment data on, for example, water quality and quantity (surface and ground water), river sediments, details of accidents and emergencies involving hazardous substances and transmitted data on forecasted passage of pollution plumes downstream, reservoir conditions, atmospheric deposition, soil quality, land cover	Good *data exchange and comparison facility*. Flexible networked *common data exchange* file format computer-based facility with data management software, which can interface with the wide range of existing databases. *Presentation* of general and basic data summaries on, for example, the impact of waste disposal, transboundary impact of pollution, mass balances and pollution loads. Provision of *data exchange*_between countries via satellite communications, networked databases, radio and telemetry links, various computer media including Diskette, CD-ROM (possibly), action programmes and published annual or other digests of statistics.
Information support for Planning, Operations, Decision-making, Research and Public Participation	Design and operation of *models* and software routines *for data manipulation* in order to support analysis, interpretation, assessment and, in an operational situation, the forecasting of events on which action has to be taken. *and registers*_for making information available to the public.
Development, use and application of management tools; dissemination of information	*Information derived would support* management review, river management planning and operations, permitting, resource planning, decision-making etc. and applied research.

Table 2

DANUBE ALARM MODEL The model was conceived to support the decision-making role of the Principle International Alarm Centres (PIAC) of the Accident Emergency Warning System. These Centres, one in each country (2 are planned in Ukraine), will have the responsibility to inform downstream countries on possible inflows of sudden pollution in the event of an accidental spillage of hazardous substances. The Alarm system is based on a 1 - dimensional analytical approach - for reasons of performance, simplicity in use and robustness. The river is divided for the purpose into flow - homogeneous reaches. In the model design, the mathematical basis relates to a spill assumed to occur instantaneously in a river with homogeneous geometrical and hydraulic characteristics. Input data consists of: - information about the spill incident, - information on the river hydrology at the time of spill, - information that defines the cross - section geometry and hydraulics, - values of parameters on dispersion and dead zones in river morphology. The model is dependent on measured data. Calibration will be important. Once results through some initial tracer experiments are considered satisfactory there will be a continual activity to refine results. The work is being funded using Applied Research funds allocated to the Programme from the European Commission's PHARE programme.
DEFF (Data Exchange File Format facility) A software - based protocol comprised of 5 related tables of information - monitoring points; analyses; remarks; methods; and determinands. DEFF has been designed to facilitate the transfer of water data internationally according to the experience and capabilities of Danube countries. The approach is complementary to the National Information Systems present in each country where operational responsibility will remain as they are - developing and changing. Danube countries will be free to decide how they wish to use DEFF alongside their existing systems; it is designed to be simple to operate and to hold past data easily. A user manual (in draft at present) is available. For DEFF to be effective, National Information Centres are being identified. These Centres will have responsibility for data and information exchange appropriate to international needs under the monitoring strategy. To implement DEFF each country will be developing two software interfaces. One that can select data from its own databases and store in DEFF and another that can extract data from DEFF and store in its own National Information System.
DEMDESS (Danube Emissions Management Decision Support System) A computerised system designed to support decision -making for water pollution control. Development was initiated as part of the USAID programme of aid to Danube countries and was commissioned in Bulgaria, Hungary, Romania and Slovakia. The latest status report of DEMDESS suggests that it may not be operational in Romania. Components of DEMDESS consist of databases on emissions, waste water treatment, regulations on water pollution control, details of the river network and ambient water quality data. The system is designed to support full evaluation of the effects of decision options and operates by integrating data and information from respective files. The last development phase of DEMDESS in 1994 was concerned with making the system more user-friendly, and enhancing capability with the introduction of a modelling facility based on the USEPA "Qual2 e" model, an emissions policy model facility and one which will enhance examination of water quality.

Table 2 contd.

NUTRIENT BALANCES

A current Applied research project commissioned in the Programme has an objective to provide nutrient balances for different past and future horizons for most of the basin. The project will produce a harmonised assessment of the nutrients production in the Danube environment (sources and sinks). Use of the information so produced will support more effective decisions on options for reducing nutrients and the impact of eutrophication, which is regarded as one of the more challenging concerns in the Danube basin.

The aim of materials accounting is to quantify material flows and stocks in a defined system and to point out their origin, destination and residual. The activity will allow a determination of nutrients that enter, stay and leave a system within a timescale. Publication of the work and the analysis that results will indicate where particular effort must be applied to reduce inputs of nutrients to the Danube surface waters.

GIS (Geographical Information System)

No paper on Information exchange would be complete without reference to GIS. The Programme has not developed a GIS for the very reason that funding has not been forthcoming for such an activity. However, a Workshop was held on this topic in 1994 and it became clear that GIS applications varied considerably in the basin. Some countries were well advanced in applications; some had no facility at all.

As a result, the conclusion of the 1994 workshop was to define an activity which would provide each country with material on which to build a harmonised facility and to be able to produce a basic map. The result is the DEGIS proposal. Earlier work in the Programme had been initiated by Equipe Cousteau which had recommended a project to produce an Environmental Decision Support System based on a Geographical Information System.

In the DEGIS debate, the objectives and benefits for GIS in support of the Programme were identified to be;

- more economical performance of projects. Availability of common base maps in digital format will save considerable effort in the initial set-up for each project and encourage inter-project exchange of information,
- regional consistency. Basin-wide mapping will introduce standardisation of data accuracy and resolution
- thematic consistency. Presentation of thematic information will allow different groups to access consistent thematic information,
- support for improved water and environmental management.

A possibility may exist to marshall CORINE data for Programme use if funds become available.

THE WORK OF THE MONITORING, LABORATORY AND INFORMATION MANAGEMENT SUB-GROUP (MLIM-SG)

Dr TIM LACK
WRc plc (Water Research centre)Marlow, Bucks, SL7 2HD
United Kingdom

KEYWORDS / ABSTRACT: monitoring / informatics / water quality / Danube River Basin / water management

A consortium, known as WTV, was appointed by the European Commission to evaluate existing monitoring, laboratory analysis and information management capabilities of the riparian countries of the Danube River Basin. Major recommendations are: establish a Trans-National Monitoring Network (TNMN) of representative stations for the assessment of the chemical and biological quality of the Danube and its major tributaries; establish National Reference Laboratories for the Danube in each riparian country; provide an information management system for the TNMN that supports exchange of data, statistical trend analysis and interfaces with national information systems in each Danube riparian country, and provide training for trainers on the system and its applications.

1. Background

The Environmental Programme for the Danube River Basin (EPDRB) was jointly established by the riparian states of the Danube at a meeting in Sofia in September 1991. The Sofia meeting established a broad framework for a multi-year action programme and agreed to set up a Task Force to oversee and co-ordinate the Environmental Programme. To assist in the implementation of the programme a Sub-Group, the Monitoring, Laboratory Analysis and Information Management Sub-Group (MLIM SG) was formed. One of the first tasks was to steer a Technical Assistance Project commissioned by the CEC's PHARE Operational Service (with financial support from a number of other donors such as TACIS and UNEP) under the advice of the Danube Programme Co-ordination Unit (PCU).

The broad objective of this project was to assist in bringing about improved environmental management through the strengthening of existing environmental monitoring networks, improving the comparability of sampling techniques and laboratory analyses, and by building upon or where necessary developing and implementing compatible information management systems capable of providing the needs of managers at the local level as well as permitting the exchange of information

I. Lyons Murphy (ed.), Protecting Danube River Basin Resources, 127–134.

at the international level. It was also recognised that the approach developed for the Danube Basin should provide outputs compatible with other major international river basins in Europe.

The more detailed objectives of this study were:

- To make an evaluation of the existing monitoring, laboratory analysis and information management capabilities of the Danube riparian countries.
- To make an analysis of options and recommendations for strengthening those capabilities.
- To identify the priority needs on a country by country basis and basin-wide.

The achievement of these objectives and the implementation of the recommendations will improve environmental management in each of the countries and assist in the development of a trans-national approach to the problems of water quality management in the Danube River Basin.

2. Organisation and Approach

The work was carried out by a consortium, (designated WTV), of WRc (UK), TNO (NL) and VKI/DHI (DK). The consortium team were assisted by local consultants from the following riparian states: Bulgaria, Croatia, Czech Republic, Hungary, Romania, Slovakia, Slovenia, and Ukraine. The Republic of Moldova was dealt with separately as it was a late participant in the Programme. Questionnaires on Monitoring, Laboratory Analysis and Information Management were prepared and the local consultants arranged their completion. Mission Teams comprised of Team Leaders from WTV and other specialists visited each country to make an independent assessment of current capabilities and future needs. The combination of expert knowledge from local consultants, data from questionnaires and experience from missions provided the WTV team with a powerful database upon which to base their judgements and make their recommendations.

3. Recommendations From the WTV Technical Assistance Project

3.1. MONITORING

The existing network of monitoring stations (developed for the requirements of the Bucharest Declaration) should be extended from 13 to 49 stations initially and, pending review after 2 years, to a total of 79 stations. This Trans-National Monitoring Network (TNMN) is firmly based on existing stations within each country's National Monitoring Network. Sampling for meteorological and hydrological data and water and sediment quality should be carried out. A list of determinands and the frequency of sampling was provided. There should be more emphasis on calculation of loads of contaminants and there are recommendations for improving the collection and integration of flow and water quality data. The purchase of flow measurement and bed profiling equipment was recommended

3.2. SAMPLING AND LABORATORY ANALYSIS

To improve the consistency of sampling and the quality of the data obtained there is a need for water and sediment sampling equipment and field monitoring equipment for determinands such as pH and Dissolved Oxygen. The procurement of this is underway. New laboratory equipment will be required to analyse for the determinands in the revised list. The needs have been identified on a country by country basis and it is recommended that a National Reference Laboratory be nominated to co-ordinate the sampling and analysis requirements of the TNMN. As well as capital equipment items, there is a requirement for support equipment (such as water deionisers) and consumables (such as reagents and gases). There is a need to improve Analytical Quality Control management in the laboratories and to carry out performance testing of the methods used. The feasibility and value of carrying out international check sample distributions has been proven and should continue in order to maintain an external pressure on quality improvement.

3.3. INFORMATION MANAGEMENT

Rather than specify a data and information management system to be used by all countries, the protocol by which countries should interchange data from the TNMN has been defined. This Data Exchange File Format (DEFF) will need to interface with each National Information System and the resource requirements necessary have been identified. In parallel with the National Reference Laboratories it is recommended that National Information Centres be identified and provided with appropriate hardware and software which have been specified. There is a need for an international working group to collect, collate and process the data from the TNMN working in a similar way to the Secretariat of the International Commission for the Protection of the Rhine. Indeed, the Rhine Commission could be a model for the international control of the Danube.

3.4. TRAINING

Clearly, with so many recommendations for change, there is a need for a major programme of training and the following key areas were identified:

- In the use, maintenance and calibration of the new sampling, monitoring and laboratory equipment.
- In specific chemical, radiological, microbiological and biological techniques.
- In laboratory management including AQC and health and safety.
- In basic computer use, data handling, processing and information management.
- In institutional development for senior management.

3.5. RESEARCH

To underpin the recommendations and to provide a basis for future developments a number of research topics were identified under the following headings:

- Mathematical modelling.
- Biological and chemical classification schemes basin-wide.
- Direct toxicity assessment and ecotoxicology.
- Impoundment and sedimentation studies.
- Fate and pathways of organic pollutants.
- Preservation and restoration of wetlands.
- Radiological studies.

4. Implementation

Needs were identified on a country by country basis and advice was given on the priorities. Wherever possible, detailed specifications for the procurement process were prepared and costs estimated. An overall implementation plan based on the findings of this project was drawn up and approved by the Programme Coordination Unit of the Danube Programme, the funding organisations and the MLIM SG.

5. Activities of the MLIM SG

5.1. MISSION STATEMENT

One of the earliest actions on the MLIM SG was to determine its Mission and the following Statement was agreed upon MLIM SG will:

- Establish a Trans-National Monitoring Network (TNMN) of representative sites for the chemical and biological quality assessment of the River Danube and its major tributaries.
- Establish reference laboratories in each Danube riparian country that will be responsible for the analysis of all TNMN samples, and that could act as national laboratory accreditation centres in each country.
- Provide an information management system for the TNMN that will also interface with national databases in each Danube riparian country.
- Provide training opportunities at levels appropriate to the needs of each Danube riparian country; priority areas are biological assessment using macro-invertebrates, assessment of pollutant loads in sediments, and the derivation of data-exchange information systems.

MLIM SG will achieve these prime objectives through three specialist working groups which will make recommendations for action to the Danube Programme Coordination Unit through the MLIM SG.

5.2. PROGRESS TO DATE

5.2.1. *Establishment of the Working Groups*
With the support of the MLIM SG Chairman, Mr Liviu Popescu of Romania, and the consultant to the PCU, Mr J Alan Tetlow, the following chairmen were appointed with the unanimous support of MLIM SG:

- Monitoring WG: Mr Milan Matuska (Slovakia)
- Laboratory Management WG: Dr Peter Literathy (Hungary)
- Information Management WG: Mr Ivan Milushev (Bulgaria).

The membership of the WGs was set by MLIM SG at four or five Danubian country experts selected for their expertise to address the priority topics.

Monitoring Working Group. The Monitoring WG has agreed on the representative sites constituting the first phase of the TNMN. These were largely those recommended by the WTV consultancy with some minor local amendments and the inclusion of five sites to cover the River Prut on the border of Moldova/Romania. There has been considerable debate on biological monitoring focusing on the relative merits of the Biological Monitoring Working Party (BMWP) System and the Saprobic Index (SI) system. The Monitoring WG has recommended the mandatory use of the SI system for the 1996 "trial year" of the TNMN, with the optional use for comparative purposes of the BMWP system if resources allow.

Laboratory Management Working Group. The Laboratory Management WG has finalised its work on the chemical determinands listed in the WTV Final Report and has recommended that all countries should aim to carry out this work for the main river and major tributaries. It has been recognised that some countries at present do not have the capacity to carry out some of the analyses due to lack of facilities and expertise. An example of such an important but difficult determinand is "oil in water" and a training workshop on this was held in Budapest in 1994. Similarly a workshop on pollutants in sediments and suspended solids is planned as part of the integrated programme of training (see Section 5.2.4). A position paper has been prepared and a common approach is recommended for all Danubian countries.

Analytical Quality Control (AQC) is recognised as being of crucial importance in this international monitoring programme. The LMWG has been involved in a number of international AQC schemes (e.g. EU EQUATE, QualcoDanube, EU IMEP-6, and AQUACHECK). The WG recognise that AQC needs to be implemented on two levels: a national AQC programme for country laboratories and the use of international check-sample schemes between countries. The WG also recommends that QA/QC procedures should apply to all aspects of the TNMN programme and this requires co-operation between the three WGs. It is also proposed that laboratories that fail to achieve adequate AQC should have access to the necessary staff training and to an "inspection facility" to advise on remedial actions to restore adequate AQC.

Information Management Working Group. The Information Management WG have agreed that, initially, data exchange will be via diskette but in the longer term will be by e-mail. Members have established liaison links with the MWG and LMWG to match the design of the Data Exchange File Format (DEFF) chemical method file, the DEFF sampling method file, the DEFF sampling and analysis files and the DEFF biological sampling file. Anti-virus software has been specified and the DEFF software has been produced as a test version. The DEFF software incorporates a statistical analysis and data display package (AARDVARK) to facilitate the production of reliable and comparable technical and management reports. Training on the use of the DEFF and its associated software is scheduled for mid-1996.

5.2.2. *Liaison and Support*

A number of liaison linkages have been established, including: Global Environmental Monitoring System (GEMS), the International Commission for the Protection of the Rhine (ICPR), the Rhine Basin Programme (RBP), the Black Sea Environmental Programme (BSEP), the Bucharest Declaration (BD), USAID, the European Environment Agency (EEA), in particular its European Topic Centre on Inland Waters (ETC/IW), the United Nations ECE Monitoring Programme, the International Association of Danube Water Suppliers (IAWD) and the newly constituted interim International Commission for the Protection of the Danube River (ICPDR).

Linkages are maintained with the Accident and Emergency Warning Systems WG through the PCU.

Technical support to MLIM SG is provided by Dr Tim Lack, the Project Manager of the original WTV Technical Assistance project and to the separate WGs by the original WTV experts: Monitoring - Dr Anders Lyngaard Jensen, Laboratory Management - Dr Peter Whalley and Information Management - Dr Tjaart Kuipers.

5.2.3. *Institutional Development*

Each country has identified its National Reference Laboratory and National Information Centre for the purposes of the Danube Programme. Apart from being the recipient institutes in the equipment procurement programme (see Section 5.2.4) they will be responsible for the quality of the data nationally and internationally and will be expected to act as centres of excellence within each country to promote the quality ideal, to stimulate the development of new methodologies and to train national staff as appropriate.

5.2.4. *Procurement Programme*

Consistent with the original recommendations of WTV, the PCU has supported a major procurement programme consisting of field monitoring equipment, boats and vans for sampling, laboratory analytical equipment, support equipment such as reagents, gases and glassware, computer hardware and software for laboratory AQC and for data handling and information exchange as well as basic office support equipment such as printers, photocopiers and fax machines.

5.2.5. *Integrated Training Programme*

Again, following the recommendations of WTV, the PCU has put in place an integrated programme of training operating at a number of technical and managerial levels. The training covers the following aspects:

- Sampling and field monitoring
- Analytical equipment training
- Laboratory equipment training
- Analytical techniques
- Macro-invertebrate taxonomy
- Sampling and analysis of river sediments
- Institutional development training for senior management.

5.2.6. *Applied Research Programme*

An Applied Research Programme has been initiated with the broad objective of helping to define the legal, policy, and management framework required to improve the environmental situation of the Danube river, the Danube delta and areas of the Black Sea influenced by the Danube. A number of projects are currently underway covering (amongst others) the following topics:

- Ecological risk assessment of pollution by heavy metals.
- Nutrient balances for individual countries of the Danube basin.
- Impact of radionuclides in surface waters and sediments of the lower Danube.
- Biodiversity studies.
- Quality of sediments and biomonitoring.
- Nutrient removal by wetlands, floodplains and reservoirs.
- Nutrient application in agriculture - protection of resources against diffuse nitrogen pollution.
- Water quality targets and objectives.
- Integrated surface and groundwater protection through wastewater treatment.

In this way the work of MLIM SG and its WGs is underpinned by a programme of practically oriented research carried out, by and large, by national experts within the Danube riparian countries.

6. Issues for the Future

The development of the Strategic Action Plan for the Danube River Basin began in 1992 and was approved by the Danube Task Force in 1994. The Action Plan makes a significant contribution to the international efforts to improve environmental management in the Danube Basin. The progress described above is consistent with the common strategic directions and common targets of the Action Plan. However, the actions taken at present are all in the short-term, focusing on urgent issues or building-blocks essential for achieving the medium and long-term targets.

Essential tasks to be undertaken include:

- integrated tributary river basin plans with water allocation and use permits.
- Adoption of consistent water quality objectives.
- Comprehensive inventory of emissions and pollution risk assessment.
- Legislation controlling the transportation of hazardous substances.
- Wastewater and sewerage priorities for cities, towns and villages.
- Environmentally sound and sustainable agricultural policies.
- Comprehensive system of charges for abstraction and discharge to water courses and sewers to provide a sustainable flow of revenue to support the continued monitoring of the Danube River Basin under the Polluter Pays Principle.

Finally, it should be borne in mind that a substantial part of the Danube River Basin has thus far been excluded from consideration because of civil war. With the continued peaceful rehabilitation of the former Republic of Yugoslavia there is an opportunity to complete the needs assessment started in 1993 and finally to include all countries of the Danube basin.

7. Acknowledgements

The author gratefully acknowledges the support of Mr David Rodda (Team Leader of the PCU), Mr J Alan Tetlow (Consultant to the PCU) and Mr Liviu Popescu (Chairman of the MLIM SG) in the work carried out in the Danube River Basin and leading to the production of this report. The contributions of the WTV team of experts and the team of local consultants is also gratefully recognised and acknowledged.

THE DANUBE GRANTS PROGRAM AND THE NGO ROLE IN MANAGEMENT OF THE DANUBE BASIN

ROBERT ATKINSON
Regional Environmental Center for Central and Eastern Europe.
1 Miklós tér
1035 Budapest, Hungary

The opinions expressed in this paper do not necessarily represent the policies or opinions of the Regional Environmental Center.

KEYWORDS / ABSTRACT: Regional Environmental Center / Danube Grants Program / Danube River Basin public participation

This paper covers the goals and activities of the Danube Grants Programme managed by the Regional Environmental Center for Central and Eastern Europe (REC) for the Danube Programme Co-ordination Unit, and funded through the Global Environment Facility. Using the results from this programme, it highlights the types of activity carried out by non-governmental organisations towards Danube issues and evaluates how these independent activities assist the protection of the Danube Basin, particularly where they involve a monitoring or research component.

1. The Background of the Danube Grants Program

The main components of this paper are a description of the Danube Grants Programme managed by the Regional Environmental Center for Central and Eastern Europe (REC), and a review of the lessons the non-governmental organisations (NGO) actions promoted by this programme give to defining a component of public participation in any action plan for water quality data sharing in the Danube River Basin.

1.1. REC MISSION AND BASIC PRINCIPLES

To set the scene it is necessary to outline the mission and basic principles of the REC and the Danube Programme. It is these principles which colour the activities and priorities of REC and Danube Programme objectives and actions towards the NGO community, such as the granting system described below.

The REC was established in 1990 as an independent, non-profit foundation by the United States, the European Union and Hungary. In 1995 the status of the Center was changed from a Hungarian foundation to an international organisation. The REC's

I. Lyons Murphy (ed.), Protecting Danube River Basin Resources, 135–148.

founding mission is to promote public participation in environmental decision-making, thereby assisting the central and eastern European countries in their transition to a democratic, sustainable society [1].

REC assistance is offered to non-governmental organisations, local and national governments, academic and scientific institutions, media and businesses in fifteen Central and East European countries: Albania, Bosnia-Herzegovina, Bulgaria. Croatia, Czech Republic, Estonia, Hungary, Latvia, Lithuania, Macedonia, Poland, Romania, Slovakia, Slovenia and Yugoslavia. This assistance is offered through information services and products, fellowships, working groups on regional issues and through grants. Granting through the Danube Grants Programme, and its implications, are the subject of this paper.

The principles of public participation in environmental decision-making, and its promotion, are a key aspect of the REC's activities. From the REC's manual on public participation, the foreword expands on this view: "public participation includes the right of all individuals to be informed of plans and projects which may deteriorate their environment .... The heart of this system is public participation, which necessarily includes prior information." [2]

In the text of this manual it continues: "Public participation is not only advocated by the Rio Declaration of 1992 and Agenda 21 adopted at the same conference; it is also included in the most recent international treaties covering the sea and of rivers, and relates to transfrontier pollution.

"The public is one of a nation's greatest resources for developing and implementing environmental laws and policies. They know the country's land and natural attributes more intimately than a government ever will. Their number makes them more pervasive than the largest government agency. The involvement of the public is crucial to the establishment and implementation of a fair and effective environmental protection regime.

"Significant benefits to the government can result from public involvement in the environmental protection process. Citizens have direct, immediate knowledge about environmental conditions in their communities. Encouraging them to share this knowledge with the government can create more informed government decisions and can reduce the likelihood that significant environmental impacts of proposed actions or policies be overlooked.

"The sheer number of the citizenry can also supplement scarce government monitoring, inspection, and enforcement resources, thus saving money and time. Members of the public can act as the government's 'eyes and ears' identifying and taking action against environmental threats or violations of applicable laws."

1.2. DANUBE GRANTS PROGRAMME

The Danube Environmental Programme, sponsor of the Danube Grants Programme, also realises that the public and citizens groups play an important role in the success of its work.

In the Strategic Action plan for the Danube River Basin 1995-2000 (section 1.7, paragraph 47) it states that: "Every person in the Danube Basin has several possible roles - as a consumer of goods and services; as a producer of waste at home and at their workplace; as a user of recreation facilities; and as citizens whose choices and actions express their spiritual, social, aesthetic and environmental values. NGOs are established by members of the public *to promote public and governmental awareness of environmental issues and to assist in the development of appropriate policies.*" [3]

It is this final sentence that is compatible with the described mission of the Regional Environmental Center. This acceptance of the NGO's role, in informing the public and government and having a say in policy decisions, allows NGOs to have a rightful place in the activities relating to the protection and utilisation of Danube River Basin water resources.

This statement is reinforced in the 1995 annual report of the Environmental Programme for the Danube River Basin: "National and regional NGO networks have a vital contribution because it enables the expression of a broad range of viewpoints concerning current and future environmental management of the Danube Basin. NGOs are valuable sources of independent expertise and advice on certain subjects and can act as a channel for the transfer of information between government and the public, and vice versa." [4]

So the Danube River Basin Grants Programme (Danube Grants Programme), as an initiative of the Environmental Programme for the Danube River Basin, was designed to support NGOs in their perceived role to assist the Programmes objectives. This role is a recognition of the need for public participation.

2. Origins of the Danube Grants Programme

The countries of the Danube River Basin (Germany, Austria, Czech Republic, Slovak Republic, Hungary, Slovenia, Croatia, Yugoslavia, Romania, Bulgaria, Ukraine and Moldova) have a long tradition of co-operating on water issues affecting the Danube River Basin (e.g. 1948 Belgrade Convention on Navigation). However, a need to strengthen environmental cooperation and to integrate and co-ordinate plans and projects was felt by the Danube countries following the collapse of the former regimes in the late 1980s. Therefore, representatives of the Danube countries, as well as representatives of other countries, financial partners and other interested parties met in Sofia (September, 1991) to agree on an initiative for restoring and protecting the Danube River, and established the Danube Task Force. At the first meeting of the Danube Task Force, held at the beginning of 1992, the Environmental Programme for the Danube River Basin (EPDRB) was effectively launched. [5]

At this same meeting, a work plan was agreed which set out a three-year programme comprising short-term, strategic, environmental management and institutional development activities. The Danube River Basin Grants Programme (for short: Danube Grants Programme) is a part of the Environmental Programme, and addresses especially the last two issues: environmental and institutional development activities. It is funded from the Global Environment Facility (GEF).

3. The Goal of the Danube Grants Programme

The goal of the Danube Grants Programme is to enhance the institutional capacity of local groups and NGOs working on Danube River-related issues. By helping local NGOs to take responsibility for a specific local problem, this contributes to improving the environment of the Danube River Basin as a whole.

To achieve these goals, the Danube Grants Programme supports organisations, working on improving environmental conditions in the catchment area of the Danube River Basin, by providing funding for specific projects. In 1995 there was up to USD 25,000 available for each country involved. It is hoped that during 1996 funds will be made available to continue this.

3.1. APPLICANTS AND APPLICATIONS

The Danube Grants Programme invited non-governmental, non-profit organisations and NGO networks working on environmental issues to apply for the grants of up to ECU 4,000 per year per organisation (N.B. The maximum amount of money awarded per year per organisation is independent from receiving grants through REC's own Local Grants Program).

Other organisations working on environmental issues in the Danube River Basin were also eligible, such as scientific associations, farmers organisations, co-operatives, scientific and technical institutions *provided* that they worked together with community-based NGOs or local community groups. Such partnerships were encouraged to enhance the capacity of NGOs and local community groups, and thus contribute to an exchange of experience and knowledge in research, data collection, training and education between institutes and local people. This is a key point in the realisation that NGOs have a role in data-sharing and monitoring activities.

Some basic conditions apply to applicants for a grant from the Danube Grants Programme, they must:

- be officially registered;
- have their activities mainly in the environmental field;
- have a basic organisational structure and/or constitution, and be a local, community-level or national organisation;
- possess at least one year's experience working on Danube River Basin issues;
- have demonstrated a capacity in project management, reporting, and accounting to assure effective project monitoring and evaluation.

Eligible NGOs and NGO networks were allowed to include in their applications other NGOs located in the Danube River Basin dedicated to preventing environmental degradation, or organisations concerned with the promotion of sustainable development, or NGOs that carry out education, information, policy and advocacy activities in the area of the Danube River Basin.

All applications had to be prepared according to the standard version of the REC's "Local Grants Program - Proposal Guidelines" issued by each REC Local Office (all in the language of that country).

Two meetings were held in which the Local Advisory Board of each of REC's Local Offices evaluated the applications for the Danube Grants Programme in 1995. So in each country two deadlines were set for receiving applications.

3.2. GRANTS AWARDED AND KEY RESULTS

During 1995 the programme provided grants of up to ECU 4,000 per year per organisation. Eligible organisations are NGOs (non-governmental, non-profit organisations) and NGO networks that work on environmental issues in the Danube River Basin area. In addition, other organisations working on environmental issues, such as scientific associations, farmers organisations, co-operatives, and scientific and technical institutions applied for Danube Grants, provided that they worked together with NGOs.

Projects funded have included community-based activities that demonstrate successful approaches to environmental problems of the Danube River Basin. The Danube River Basin consists of the river Danube and all the rivers and streams that eventually (so also via other rivers) flow into the Danube.

Priority project areas for the Programme are:

- Activities to build the capacities of NGOs (or the general public) and community groups in managing and conserving the ecosystems of the Danube River Basin (especially wetland, alluvial forests and other near-river habitats);
- Training programmes, exchanges of experience through networking, communications and advocacy activities;
- Research, data collection and prevention of pollution of rivers from inadequate waste disposal (urban and rural). Such activities should be linked to information dissemination, publication, awareness-raising and campaigning activities;
- Community-based local management plans and their implementation for rivers, lakes, groundwater and reservoirs influencing the pollution of international waters;
- Providing support to promote sustainable agricultural and irrigation practices to remedy damaging impacts which affect the quality and quantity of surface and groundwater systems;
- Activities to reduce excessive plant growth harmful to animal life, especially eutrophication reduction.

In Annex 2 the listing of all Danube Grants awarded, by the REC, in 1995 can be found. The following two tables (Tables 1 and 2) use data from this annex. This granting programme has been very successful in supporting the activities of NGOs working on Danube related issues. Ninety-six individual projects have been completed in 1995, and it is hoped that the programme will continue in 1996.

The awareness of the general public has been raised through campaigns, information and participatory actions. The capacity of many NGOs to develop their objectives and priorities has been heightened. Concrete and accurate data has been collected, analysed and used to develop alternative strategies. Threatened areas have been saved by NGO management projects and the status of ecologically important areas have been revealed. In some cases, groups from two or more countries are working together on a transboundary problem, something that government authorities (particularly local) often find difficult. All these activities highlight the contribution that citizens associations can give to the environmental issues concerning the Danube River and its basin.

TABLE 1 - Summary of grant amounts and numbers

COUNTRY	Number of grants awarded	Total amount of USD spent	Average amount of USD per grant
Bulgaria	11	25,000	2,273
Croatia	11	24,100	2,272
Czech Republic	8	22,100	1,934
Slovakia	16	25,000	1,563
Hungary	30	25,000	833
Romania	11	25,000	2,273
Slovenia	9	24,997	2,777
TOTALS	96	172,088	1,793

This table shows the totals of Danube Grants awarded in each country and the total funds spent, plus the average size of grant awarded. Two key points come out from Table 1. One is that Hungarian grants make up approximately one third of the total number awarded across these countries. This is not surprising considering that the entirety of Hungary is situated in the Danube River Basin and therefore all water issues relate to the programme in some way. Conversely the other interesting point is that the Czech Republic was the only country where a large portion of grants funds went unspent.

On looking into more detail on topic or issues covered by the awarded projects one can see the activities that NGOs are active in. This will show whether the public or NGO community has an interest, or even the capacity, to be involved in policy making decisions of the Danube River Basin. An assumption of the mission of the REC described above.

Using the information on the grant projects awarded the following approximation of numbers for the following three kinds of project can be found (out of a total of 96):

TABLE 2 - Project activity amounts

PROJECT ACTIVITY	NUMBER OF GRANTS
Monitoring/data component	36
Supplying information/data	20
Management of river process/species	18

This clearly shows that NGOs and community groups are actively involved in monitoring the environmental situation of the Danube, they are actively supplying information on its status to the general public, and in some cases are carrying out projects and research to manage parts of the basin's ecology.

It is perhaps proper that some of these projects are highlighted to reveal the depth of independent activity being carried out by NGOs. In Table 3 a selection of ten representative projects has been made to illustrate these activities.

TABLE 3 - Selection of indicative NGO projects from the Danube Grants Programme in 1995

Project Code	NGO name	Country	Project Title
DO6006	Borrowed Nature	Bulgaria	Appraisal of the Danube River pollution caused by waste waters discharged from Ruse
DO6017	Union for Nature Protection	Bulgaria	Monitoring of Danube Fish Diversity
DO2003	Czech Union of Nature Protectors	Czech Republic	Higher quality and cheaper water for Brno
DO3012	SZOPK - "SITNO"	Slovakia	Independent environmental monitoring for the DEP
DO3019	Association of Forest Protectors	Slovakia	Analysis of damage to the hydrological regime of the eastern Slovakian region
DO4001	Tree of Life	Hungary	Monitoring waters in Heves and Borsod counties, with 14 local groups
DO4008	Green Action Society	Hungary	Documenting illegal polluters
DO4011	Goncol Foundation for Living Waters	Hungary	Monitoring techniques information resources
DO4050	Green Heart	Hungary	Duna-chain, Tisza-line program - co-ordination of 50 monitoring groups
DO5015	The Ecological Club Transylvania	Romania	Somes '95: Involvement of local communities in the ecological surveillance of mountain rivers

This selection shows NGOs monitoring, collecting data and using it to inform the public and prepare management plans. These are real activities that make a real impact locally, and sometimes nationally.

4. Other NGO Programmes on the Danube

Outside of the activities funded by the Danube Grants Programme other NGO activities are taking place based on the use of monitoring of water quality and biological diversity. The World Wide Fund for Nature (WWF) runs its own Green Danube Programme and develops numerous projects to help protect and restore the natural Danube habitats [6].

One such project, entitled: Expanding the Capacity of NGOs to Monitor, Conserve and Rehabilitate Danube Area Wetlands, shows the faith in and determination to use NGOs to contribute to preservation of the Danube ecosystems. It recognises that often NGOs are the only people who have the skills, speed and commitment to develop new activities for conservation. These are areas where the government does not have the capacity, or more usually the resources, to act.

Nadacia SOSNA, based in Kosice, Slovakia, shows how an east European group can develop and manage a complex cross border programme. SOSNA is the main co-ordinator of the Water Monitoring Project between Kosice and Miskolc. There are several NGOs in Kosice, Slovakia, and Miskolc, Hungary, with expertise in water protection and management.

The project consists of monitoring water in and around the two cities on the Hornad River (a Danube tributary). Due to the presence of heavy industrial activities, the water quality and natural environment of this area has been significantly impaired.

The NGO highlights the lack of public understanding, the few organisation (state or NGO) which have devoted resources to educate and inform the public in the area of water resource conservation and the lack of willingness by the local governments, in both states, to solve the problem of poor water conservation and water contamination [7].

They hope to alter this situation by: informing the public through publication of water monitoring results in the media, through exhibitions and publications; by improving the relationship between the participating NGOs and local government environmental offices; and through facilitating discussion about the problems and their solutions in areas of environmental concern.

In all, ten Slovak groups will work on monitoring a section of the River Hornad from Kysak to Zdana (near Kosice), and Seven Hungarian groups from Zsujta (at the border) to Gesztely. As well as water quality monitoring they will look at the fauna and flora and begin conservation activities. These two projects highlight how NGOs can and are making a difference.

6. The Message and Conclusions

So what is the message of these NGO activities to an action plan for data sharing in the Danube River Basin? It is clear that whether the grants from the REC are there or not many of these activities would still be happening. However, the ease of carrying them out would be very seriously hindered. Also the willingness of the Danube Environmental Programme to support this programme shows the recognised need for NGO involvement in the Danube process.

It is also clear that NGOs are capable of generating valuable, independent data, and of utilising this data to inform the public and press for changes to policy. This is an element that needs to be included in any action plan for continuing and improving sharing of water quality data. It now remains to develop forms of information for the public (general and informed) that is unbiased and non-interpreted, and based on sound accurate data.

From the REC's manual on public participation the following table (table 4) summarises the areas where information resources are a precondition for public participation:

TABLE 4 - Preconditions for public participation, information resources

Methods	Mechanisms	Constraints	Tasks
Public/NGOs are guaranteed access to already existing government information on environmental issues	Statute granting general access to government information (with procedures)	Traditional limits on legal access to information	Develop and lobby for broad access to information statute with limited exemptions and specific procedures
Public/NGOs are guaranteed access to already existing industry data on environmental compliance	More limited statute allowing access only to government's environmental information	Political opposition	Develop and lobby for self-monitoring and reporting requirements to be included in environmental laws
Government (including local) collects and maintains publicly available environmental information (through self-monitoring and reporting, permitting, and EIA requirements, for example	Requirement that governments develop and disseminate environmental information and make it available at convenient locations	Unwillingness of industry to cooperate	Develop requirements for governments to publicise environmental data
Industry collects and maintains publicly available environmental information	Annual governmental report on the state of the environment	Public lethargy/disinterest in obtaining and using information	Develop requirements for governments to incorporate data gathering and dissemination requirement into appropriate laws (e.g. administrative law, planning act, environmental law, EIA law)
Public/NGOs collect and publicise data themselves, with the help of universities, students, specialists, etc	Government reporting of environmental accidents	Traditionally broad definition of state secrets	Establish public right to obtain information without proving direct need
Public/NGOs/government publish information leaflets	Requirement that polluters disclose information they already have to the public on request	Need to protect trade secrets and privacy where appropriate	Ensure that environmental databases are compatible with access to information rules
Media publicise environmental information	Requirement for industry self-monitoring and reporting, with public access to reports (as under U.S. Clean Water Act		Encourage NGOs to increase information gathering and dissemination activities
Public/NGOs hold hearings (official/unofficial) on environmental issues)	Requirement that industries develop and disclose information concerning use and emissions of toxins		Establish advisory committees
Public/NGOs serve on advisory committees and report on activities	National environmental monitoring database system		Establish ombudsman

TABLE 4 - Continued

Methods	Mechanisms	Constraints	Tasks
Environmental ombudsman obtains information as representative of the public and regularly publishes summaries of information	Public hearing procedures		Train journalists and other media representatives in environmental issues
	Broad access to the government and industry information given to ombudsman and/or advisory committees Legal enforcement of principles of equal right of access and non-discrimination Right to judicial review		

The table shows what kind of information (data) should be available and how this may be put into practice. Plus, of course, any constraints to this implementation. One point that relates directly to any action plan on data sharing is that we should "ensure that environmental data bases are compatible with access to information rules." Also this table shows the point which is stressed throughout this paper that: "NGOs collect and publicise data themselves, with the help of universities, students and specialists." From the above described grants programme it can be seen that, encouragingly, this is already happening concerning the Danube River Basin.

Recently the United Nations Economic Commission for Europe prepared a draft Convention on access to environmental information and public participation in environmental decision-making. This is being carried out by the Committee for Environmental Policy in accordance with the relevant decisions included in the Ministerial Declaration adopted at the "Environment for Europe" conference in Sofia 25 October 1995. Draft elements of the Convention on Access to Environmental Information and Public Participation in Environmental Decision-Making (dated 11 April 1996) have been prepared and it is planned that the convention will be ready by the time of the next "Environment for Europe" Conference in Copenhagen during 1998. This will apply to a right to information on water quality [8].

It is now obvious that the involvement of NGOs (the informed public) and the general public in sharing of data is not a luxury but a basic right. Their involvement is an activity and requirement of those collecting and sharing data. It needs to be stressed that, as can be seen from the above NGO projects, that NGOs will find out environmental data and they will inform the public - in central and eastern Europe they may not be so effective as in the west till now, but they are certainly becoming so.

It is also obvious that NGOs can help the government to improve the environmental situation of the Danube. Provided they are given a fair place on the stage. They can help to cover areas of environmental concern or projects for environmental improvement where governmental authorities may not reach. The shear diversity of NGOs is their major strength. The Danube Environmental Forum, a coalition of Danube interested NGOs supported by the Danube PCU, has set up two NGO

Information Centers (one in the upper river, Brno, and one in the lower Danube, Bucharest). The objectives of these centers are:

- **to stimulate full public participation in the environmental management of the Danube River Basin;**
- **to act as focal points between the regions NGO community and communication link to other information centers, the PCU and governmental bodies;**
- **to develop an environmental network to identify project areas for which grants can be attracted.**

At the very least any action plan for improved and continued data sharing should incorporate these NGO information centers as a first step.

ANNEX 1

Information on organisations and programmes.

The management of the Danube Grants Programme is as follows:

The Danube Grants Programme is a part of the Environmental Programme for the Danube River Basin. The funding for the Danube Grants comes from the Global Environment Facility (GEF), through United Nations Office for Project Services (UNOPS).

The co-ordinating body of the Environmental Programme for the Danube River Basin is the Danube Programme Co-ordinating Unit (PCU), which is located in Vienna.

On the advise of the PCU, UNOPS has contracted out the administration and management of the Danube Grants Programme in Czech Republic, Slovak Republic, Hungary, Slovenia and Croatia to the Regional Environmental Center for Central and Eastern Europe (REC), which has Local Offices in each of these countries.

The Regional Environmental Center for Central and Eastern Europe

The Regional Environmental Center for Central and Eastern Europe (REC) is an independent, non-advocacy, non-profit foundation, established in 1990.

The REC's mission is to promote co-operation among diverse environmental groups and interests in Central and Eastern Europe (CEE); to act as a catalyst for developing solutions to environmental problems in this region; and to promote the development of a civil society. Beneficiary countries are Albania, Bosnia-Herzegovina, Bulgaria, Croatia, Czech Republic, Estonia, Hungary, Latvia, Lithuania, Macedonia. Poland, Romania, Slovak Republic, Slovenia, and Yugoslavia. The REC's headquarters is in Hungary, and in most of the beneficiary countries there are Local Offices.

The REC has developed several programs to achieve its mission, such as grants programs for non-governmental organisations (NGOs), an information service available to all organisations, fellowship programs to enhance the individual capacities of people working on environmental issues in CEE, and special projects which are mainly targeted towards local and national governments.

Environmental Programme for the Danube River Basin

The objective of the Programme is to establish an operational and regional basis for strategic and integrated environmental management. The co-ordinated approach taken seeks to protect and enhance regional and transboundary environmental value, integrate environmental policy, involve public participation and promote action in both the public and private sectors.

The main Programme activities are: Accident Emergency Warning System; Institutional Strengthening to improve management and operational activities;

integrated regional environmental studies; national reviews of the status of the water environment; monitoring and laboratory management; producing a Strategic Action Plan; implementing and Applied Research Programme; Data Management to support the exchange of textual information within the Programme and the Danube Grants Programme to enhance the institutional capacity of NGOs working on Danube River-related environmental issues.

Danube Programme Co-ordination Unit

To support the implementation of the Environmental Programme for the Danube River Basin the activities of the Danube Task Force, the Danube Programme Co-ordination Unit (PCU) was set up in July, 1992. The responsibilities of the PCU are to monitor and implement Programme activities; facilitate information sharing; establish, maintain and support the national focal points, plan and co-ordinate Danube Task Force meetings; facilitate G-24 involvement; and report quarterly and annually to Danube Task Force members.

Global Environment Facility

The Global Environment Facility (GEF) is a programme aimed at assisting countries to contribute towards solving environmental problems. The GEF provides grants for mainly technical assistance and investment projects. Responsibility for implementing GEF projects is shared by UNDP (United Nations Development Programme), the World Bank and UNEP (United Nations Environmental Programme).

References:

1. Annual Report 1995 - The Road to Sofia. (1996). *The Regional Environmental Center for Central and Eastern Europe*. Budapest, Hungary. 32 pages.
2. Manual on Public Participation in Environmental Decision-Making - Current Practice in Central and Eastern Europe. (1994). *The Regional Environmental Center for Central and Eastern Europe. Budapest*, Hungary. 365 pages.
3. Strategic Action Plan for the Danube River Basin 1995-2005. (1995). *Environmental Programme for the Danube River Basin*. Vienna, Austria. 109 pages.
4. Environmental Programme for the Danube River Basin - Annual Report 1995. (1996). *Danube Programme Coordination Unit*. Vienna, Austria. 24 pages.
5. Danube Grants Programme - Announcement Brochure. (1995). *The Regional Environmental Center for Central and Eastern Europe*. Budapest, Hungary. 4 pages.
6. Green Danube Programme Brochure. (n.d.). *World Wide Fund for Nature International*. Vienna, Austria. 8 pages.

7. Water Monitoring Project between Kosice and Miskolc Leaflet. (n.d.) *Nadacia SOSNA*. Kosice, Slovakia. 6 pages.
8. Draft Convention on Access to Environmental Information and Public Participation in Environmental Decision-making. (April 1996). *United Nations Economic Commission for Europe*. Geneva, Switzerland. 22 pages.

TRANSBOUNDARY POLLUTION REDUCTION IN THREE DANUBE RIVER BASINS

RON ORACH
Metcalf & Eddy, International
30 Harvard Mill Square
Wakefield, Massachusetts 01880

KEYWORDS / ABSTRACT: water quality monitoring / Danube River Basin / transboundary pollution / US Agency for International Development

Metcalf & Eddy International, Inc. is working on a project to detect and reduce transboundary pollution sources on Danube River tributaries and to minimise the effects of these sources on downstream water users. The project will implement improvement in Hungary, Slovakia, and Romania, providing equipment, technical assistance, and training in connection with a variety of discrete tasks in each country that are designed to either prevent/control sources of pollution or to mitigate the downstream impacts of pollutants. The project is one of implementation rather than study and includes a strong emphasis on supply and installation of equipment.

1. Project background

1.1. PURPOSE

Metcalf & Eddy International, Inc. (M&E) is working under contract with the United States Agency for International Development (USAID) to implement a project known as the Transboundary Pollution Reduction in Three Danube River Basins Project. The project's ultimate objective is to reduce the emission of toxic compounds and pathogens to international waters from industrial and municipal sources. Of particular concern is the protection of waters which are used for potable or irrigation water supply and those used for fishing or bathing. The project is designed to meet the GEF criteria for international waters and was jointly developed in close co-operation with the governments of Slovakia, Romania, and Hungary and other international donors.

The threat to human health by exposure to polluted water in the project areas will be reduced by:

1. reducing direct and indirect discharges of toxics from industries,
2. reducing discharges of toxics and pathogens from municipal wastewater treatment plants, and

I. Lyons Murphy (ed.), Protecting Danube River Basin Resources, 149–164.

3. developing the capability to monitor and to quickly respond to periodic spikes of contamination in potable and irrigation water uses.

The project is being jointly funded by the Global Environmental Facility (GEF), USAID, and the host countries. USAID is planning to contribute up to $6.5 million to the GEF for the Danube project. Improvements due to these investments are intended to be measurable by the end of the project's three-year life.

1.2. GLOBAL ENVIRONMENTAL FACILITY

The Global Environmental Facility (GEF) was created in late 1990 to provide new and additional financial resources to address global environmental issues in four areas: reduction of greenhouse gas emission, protection of biodiversity, protection of international waters, and reduction of ozone layer depletion, The GEF is managed by three implementing agencies- the World Bank, the United Nations Development Program (UNDP), and the United Nations Environment Program (UNEP), in accordance with a Tripartite Agreement that draws on the respective strengths of each agency.

Funds are contributed to the GEF through parallel and co-financing arrangements and contributions to the GEF core fund. Since 1990, industrialised and developing countries have pledged approximately US $1.2 billion to support the program activities during the GEF's three-year pilot phase.

The US pledged $1.5 million to the GEF in parallel-financed USAID projects to the initial three year Pilot Phase of the GEF. As a result, USAID has developed GEF projects world-wide for contribution to the GEF. While USAID retains management and financial controls, all proposed GEF projects must meet GEF criteria and pass the GEF approval process.

The GEF classifies as "international waters" those ground and surface waters which more than one nation has access to or makes use of. The GEF recognises that the environmental integrity of many international waters is threatened by pollution, which impairs the biological productivity of the waters, and to rates of withdrawal from surface and groundwater systems exceeding their recharge rates. Municipal and industrial waste discharges and poor watershed management practices through both point and nonpoint sources of pollution are threats to downstream water users. GEF also recognises that the economic and ecological functioning of international waters is important to human well being and security, social and economic development, land fertility, and maintenance of biological diversity.

GEF projects in the international waters category must meet the general GEF criteria, be cost-effective and must be designed for sustainability. In addition. projects should meet at least two of the following criteria (The USAID-GEF project meets the first four):

- Development of basin-wide monitoring and inventories of pollution sources,
- Management plans for pollution reduction, integrating basin land use, water extraction, waste and sediment flows;

- Interventions to improve institutional capability to enhance human well-being, especially through maintaining and enhancing the productivity of water ecosystems;
- Sustainable maintenance of water-based biodiversity,
- Adaptation of local water-based ecosystems to global warming in order to protect water systems and than biodiversity,
- Monitoring and assessment of port and traffic-based pollutants and management plans for their reduction.

The project will generate regional benefits and reduce transboundary pollution of international waters by:

1. reducing direct and indirect discharges of toxic chemicals through waste minimisation and improved treatment of wastes at industries,
2. reducing discharges of toxic pollutants and pathogens through movements in municipal wastewater treatment IMP plants, and/or
3. providing the ability to monitor accidental and chronic cross-border contamination and developing strategies to respond to periodic spikes, of contamination through modification of Irrigation and/or drinking water withdrawals or improved treatment.

Each of the interventions has also been selected to include those agencies and industries with expressed and/or observed commitment to environmental improvement and the ability to meet the long-term costs of the changes initiated by the project.

2.1. GENERAL WATER QUALITY OF THE DANUBE AND ITS TRIBUTARY BASINS

Water quality problems are generally more severe in the tributaries than in the main stem of the Danube River. The higher flow rate of the main stern rapidly dilutes toxic materials and assimilates oxygen demanding substances, so there are few significant dissolved oxygen sags or toxic hot spots in the main stem. Also, a large volume of sediment is transported along various parts of the main stern before being intercepted by a number of major reservoirs. This sediment scavenges toxic substances from the stream within short distances downstream of introduction. Studies were directed to those Danube tributary basins which appeared to have the most severe water pollution problems.

This project includes three of those tributary basins: the Hornad/Hernád, the Bodrog/Upper Tisza, and the Cris/Körös. Each basin has major water pollution hot spots located near an international border, where any transboundary contamination upstream of the border could affect the downstream nation's population.

2. 1. HORNAD/HERNÁD AND BODROG/UPPER TISZA RIVERS

The Hornad/Hernád River originates in Slovakia and flows into the Tisza River after crossing into Hungary. Part of the Bodrog and Upper Tim Rivers originate in the

Ukraine, Slovakia, and Romania; the Tisza then flows through the plains of Hungary before reaching the main stem of the Danube River.

A basin study of the Hornad/Hernád prepared for USAID's Water and Sanitation for Health (WASH) Project, identified a number of hot spots in the basin and suggested possible interventions. According to the WASH study, major sources of pollution (or hot spots) in the Hornad/Hernád River Basin in Slovakia include four severely overloaded municipal wastewater treatment plants, an iron and steel works with serious phenolic discharges, and various discharges of mercury and other heavy metals including a copper smelter, a sludge lagoon, and other past and current mining activities. A basin study of the Bodrog/Upper Tisza prepared for the European Bank for Reconstruction and Development (EBRD) by the Danish firm, Carl Bro, identified a number of hot spots in each of the nations of the basin and suggested possible types of interventions. According to the Carl Bro study, major sources of pollution (or hot spots) in the Bodrog River Basin and its tributaries in Slovakia include two severely overloaded municipal wastewater plants, contamination from large industries (including chemical production, textile productions and wood processing), a variety of high chemical-oxygen-demand (COD) wastes from small to mid-sized factories, and problems with leaking chemical disposal areas and contaminated sites.

As with most of the diagnostic studies prepared for the Danube River basin, the portion of these studies addressing the Hungarian parts of the Hornad/Hernád and Bodrog/Upper Tisza Basins presented little information documenting the health impacts of upstream sources on Hungary. This is understandable because such data are extremely limited, their quality is sometimes questionable, and the basin studies have focused most directly on identifying sources of contamination and ways of reducing this contamination within each country. For this reason, activities in Hungary were designed to respond to potential impacts from upstream contamination as identified during a USAID team visit to Slovakia and Hungary in January 1994. This team met with local and national representatives in each country and identified host country concerns and supporting data. Hornad/Hernád and Bodrog/Upper Tisza project tasks are based on the WASH and Carl Bro studies and recent USAID mission findings.

Activities in the Hornad/Hernád and the Bodrog and their tributaries in Slovakia aim to reduce pathogenic and phenolic contamination downstream of the failing Kosice municipal wastewater treatment plant and to assist the Slovak government in monitoring for spills of oil products and other toxic substances in tributaries of the Bodrog originating in Ukraine. This monitoring will be paired with measures to respond to contaminant spikes by modifying bank-fed well usage and/or treatment technologies that are new at the Ukraine/Slovak border.

Activities in the Hornad/Hernád and Bodrog/Upper Tisza basin in Hungary focus on assistance to the Hungarian government in monitoring for transboundary contaminants and developing measures to respond to contaminant spikes such as temporarily blocking or protecting intakes from the Upper Tisza to the East Main Canal and/or the water intakes for the city of Debrecen from this canal.

2.2. CRIS/KÖRÖS RIVERS

The various branches of the Cris and Körös Rivers originate in the mountains of western Romania and flow into the Hungarian plain before reaching the Tisza River and eventually the Danube. Basin studies of these streams were prepared for the World Bank, using GEF funds, by the German/French consortium, Lahmeyer International/BCEOM. One study addressed problems in the Barcau River as well as the Rapid, Black, and White Cris Rivers, all in Romania. The companion study addressed these rivers in Hungary, where they are known respectively as the Berettyó River and the Rapid, Black, and White Körös Rivers. Within Hungary, the Black and White Körös combine to form the Double Körös. The addition of the Rapid Körös then creates the Triple Körös, which flows into the Tisza River, a main tributary of the Danube River.

Major sources of pollution in the Romanian headwater of the basin include three areas, all identified as hot spots by the Lahmeyer International BCEOM basin study:

1. oil extraction to the north, which pollutes the Barcau/Berettyó River,
2. a broad variety of industrial sources in Oradea in the central portion of the basin polluting the Rapid Cris/Körös River through direct discharges as well as indirect discharges to the city's overloaded wastewater treatment plant, and
3. uranium and non-ferrous metals mining in the southern portion of the basin affecting the Black and White Cris/Körös Rivers.

Similar to the work done on the Hornad/Hernád and Bodrog/Upper Tisza basins, the portion of the Lahmeyer/BCEOM study which addressed the Hungarian parts of the Cris/Körös basin presented little documentation of the effects of upstream pollution sources on health in Hungary. Again, such data are extremely limited and their quality is sometimes questionable. In addition. the previous basin studies have focused most directly on identifying sources and ways of reducing contamination within each country. Therefore, activities in Hungary under this project are designed to respond to potential impacts from upstream contamination as identified during the previously noted USAID mission of January 1994. The team met with local and national representatives in Slovakia, Hungary, and Romania and identified concerns and supporting data. The project tasks for the Cris/Körös River Basins are based on the Lahmeyer/BCEOM studies and recent USAID mission findings.

Activities in the Romanian portions of the Cris/Körös basin address three major areas of contamination. In the north, contamination from oil extraction and processing will be addressed by treating phenolic wastes not removed by existing facilities and by designing a treatment systems for oily runoff from active and inactive extraction and processing areas. Assistance to the Romanian government in this area will improve monitoring to permit better enforcement of discharge regulations and the rapid transmission of accident warning information to affected downstream users in both Romania and Hungary.

In the central portion of the basin activities concentrate on the town of Oradea. Recommendations include addressing contamination from direct and indirect discharges from two sister companies of SINTEZA Chemicals. implementing a pollution prevention outreach program for ten mid-sized metalworking plants, and improving the city's wastewater treatment facility. As in the northern part of the basin, assistance to the Romanian government is included to permit better enforcement of discharge regulations and the rapid transmission of accident warning information.

In the southern basin, interventions address mining activities: metals contamination from non-ferrous mining and contamination from abandoned and active uranium mines, both affecting the Black Cris/Körös River.

The rationale for activities in the Hungarian portions of the Cris/Körös centre around cross-border monitoring and the development of protection strategies for possible contamination of fish farming, irrigation withdrawals, and certain areas of ground water extraction.

3. Common project elements

3.1. DEVELOPMENT OF MLIM/AEWS STATIONS AND PROCUREMENT OF EQUIPMENT FOR THEM

The steps being used to develop the MLIM/AEWS stations in the project area are:

- Determine which information is to be uniformly gathered and reported to MLIM and AEWS
- Select physical locations for each of the sampling/early warning stations to provide ready and safe access for inspectorate personnel, security from vandalism and from flood damage, and a source of electrical power, if needed
- Determine what information can be remotely obtained/analysed/transmitted and what equipment is needed to do this.
- Determine which analyses can be done in the field and what types of portable equipment are needed to do these analyses.
- Determine what analyses will have to be done in a central laboratory and what pieces of laboratory equipment will be needed to do these.
- For each inspectorate involved in MLIM/AEWS station set-up and operation determine what pieces of equipment may be already available to do any of the various sampling and analytical procedures.
- From each inspectorate, get a list of the equipment suppliers which they have used recently and an assessment of the quality, reliability, and serviceability of the equipment. An important factor in assessing suitability of suppliers is the availability and annual cost of spare parts and consumable items.
- Contact a selected group of suppliers to obtain budget prices for equipment and supplies.
- Combine budget prices with quantities of equipment for each inspectorate to
- determine the estimated cost of purchasing the equipment needed for the program. Prepare a priority list of the equipment that has been requested. If the lowest acceptable bids for equipment should exceed the available equipment budget, this list will help inspectorates maximise the purchase of the most important equipment

Based on responses from the inspectorates, prepare a list of equipment manufacturers/suppliers who will be invited to bid on the equipment.

- Advertise the equipment tenders using the specifications published in the February 1994 MLIM Final Report (Volume V).

Because of the uncertainty of the timing of purchases due to government approval of Memoranda of Understanding (MOU's) in the three countries, there will be a provision made in which the equipment will be purchased for one, two or all three countries in one, two or more lots. The equipment supplier will be required to hold firm prices for some period to allow for the possibility of multiple purchases, The equipment for each country will be consigned to one central authority which may be the Ministry of the Environment or Industry, for example, who would clear the equipment through customs and distribute it to the proper inspectorates.

The actual field set-up and installation of equipment will be the responsibility of the supplier to insure that the equipment performs as specified. The supplier will also be responsible for providing training on the use and routine maintenance of the equipment.

A three-year supply of spare parts and consumable items (hoses, filters, reagents, etc.) will be provided for each piece of equipment.

3.2. TRAINING

Local inspectorate staff will be trained in the following areas, as needed:

- methods of sample collection operation and maintenance of sampling and analytical equipment use of analytical methods execution of the emergency response plan
- municipal wastewater treatment plant staff will be trained in the following areas, as needed
- treatment process monitoring and control
- quality assurance/quality control (QA/QC)
- data management treatment facility operation and maintenance

Industrial officials and staff will be trained in the following areas, as needed:

- industrial treatment facility operation and maintenance
- treatment and production process monitoring and control
- industrial pre-treatment facility operation and maintenance

Mining personnel will be trained in the following areas, as needed:

- wastewater treatment operation and maintenance
- treatment process monitoring and control and discharge monitoring.

3.3. EQUIPMENT AND CONSTRUCTION

M&E will provide the following equipment to the project through competitive bids:

- composite, grab. and continuous sampling equipment;
- analytical instruments and materials for field and possibly continuous testing of dissolved oxygen, pH temperature, turbidity, and oil;
- communication and computing equipment (radios, faxes, phones, computers, software, etc.) to implement emergency response plans;
- spill containment equipment (e.g., floating booms or boats),
- wastewater process equipment,
- sampling stations for point source discharges, in stream sampling,
- portable analytical capability for phenol and petroleum fractions, BOD, and COD,
- equipment to augment/enhance the existing carbon adsorption and biological degradation treatment systems to better control phenol emissions,
- water de-ioniser, GC, AAS, balance, and composite sampler,
- locally-supplied construction necessary for lagoon system modifications,
- oil separation and handling equipment,
- minor process or laboratory equipment,
- process control instrumentation for chromium treatment and equipment for treating zinc,
- equipment for the removal of phenol from wastewater,
- process control instruments and laboratory equipment,
- treatment process equipment for the removal of uranium from mine drainage ore,
- separation equipment to reclaim valuable material from tailings,
- surface and groundwater monitoring equipment,
- pH adjustment, precipitation, and cyanide destruction equipment.

3.4. TECHNICAL ASSISTANCE

M&E will provide technical assistance to national and local government representatives to ensure that:

- appropriate sampling and analytical methods arc finalised in accordance with AEWS and MLIM programs;
- a rapid response plan is developed, and
- a communication network is established which builds on existing systems and the work of the AEWS and MLIM programs.
- a hydraulic analysis of stream travel time between sampling points and points of water withdrawals in order to plan emergency response timing to contamination incidents.

M&E will provide technical assistance to petroleum production and refinery personnel in waste minimisation and water reuse techniques and treatment process modification design and pilot testing to reduce emissions. Assistance in managing runoff will include hydrologic and geotechnical evaluation in addition to development of a passive runoff treatment process design using constructed wetlands and other techniques. This phase will also include pilot testing.

M&E will provide technical assistance to municipal and industrial wastewater treatment personnel in sludge treatment process optimisation and industrial wage pre-treatment program design, including control of heavy metals. M&E will also provide technical assistance to municipal officials in sludge drying process optimisation and introduction of alternative sludge drying and disposal technologies.

M&E will provide technical assistance necessary for tertiary treatment process design, and pilot testing and development of O&M procedures.

M&E will provide technical assistance in determining the appropriate oil and grease removal upgrades that are necessary for the municipal wastewater treatment plant.

M&E will conduct an O&M audit, identify and implement low-cost process modifications, and design an industrial waste pre-treatment program, including control of heavy metals and toxic organics.

M&E will provide technical assistance in waste minimisation and treatment project design and pilot testing.

M&E will provide technical assistance in pre-treatment process design and pilot testing.

M&E will provide technical assistance to industrial representatives that will include technology transfer through waste minimisation audits and treatment process evaluation.

For ongoing operations, M&E will provide technical assistance to mine personnel in waste minimisation in the mining operation and treatment process design and pilot testing. Under the second activity, in old ratings areas, hydrologic and geotechnical investigations will be performed to design site closure and runoff control measures. Monitoring and maintenance systems will be set up. Hydrologic and geotechnical investigations necessary as a basis for tailings pond redesign will be conducted.

4. Project task descriptions

Early detection of contamination in surface water as it crosses into Hungary from Slovakia or Romania, and into Slovakia from the Ukraine will allow for timely notification to downstream users to minimise impacts. Responses by users include cessation of irrigation and drinking water withdrawals and spill containment measures. Cross-border monitoring facilities will be compatible with the approved recommendations of the Task Force Subgroups for Accident Emergency Warning System (AEWS) and Monitoring, Laboratory, and Information Management MLIM) program now being implemented and will assist in the implementation of these programs throughout the entire Danube drainage basin.

Improvement in the ability of responsible agencies to monitor the Bodrog and Hornad/Hernád Rivers at the Slovak border: the Szamos, White, Black, and Rapid Cris/Körös and Samos Rivers at the Romanian border, and the Uh River at the Ukrainian/Slovak border for pathogens and toxic substances (including phenols, formaldehyde, metals, oil, and grease, and radioactivity) will permit accident warnings to be issued and temporary modifications to be made to downstream water usage. resulting in a reduction of toxics reaching downstream water users. Surface water uses potentially affected include irrigation of food crops, bathing and swimming; sport and subsistence fishing; and bank-fed recharge of shallow potable wells in rural areas.

The following is an overall summary of the various tasks by country.

4.1. HUNGARY: SEVEN TASKS

4.1.1. *Task H1. Slovakia/Hungary Cross-Border Monitoring*

This task will provide equipment technical assistance and training to local authorities to monitor pollutants of concern where the Bodrog/Upper Tisza Rivers and the Hornad/Hernád Rivers cross the border and to take appropriate actions to mitigate downstream impacts.

Sampling will be conducted at two sites: one on the Hornad/Hernád and one on the Bodrog, both at the Slovak border.

4.1.2. *Task H2. Eastern Main Canal/Debrecen Drinking Water Protection*

This task will provide equipment, technical assistance, and training to local authorities to monitor pollutants of concern at the confluence of the Bodrog and Tisza Rivers at Tokaj and to take appropriate actions to minimise impacts on the drinking water supply for the city of Debrecen.

The border monitoring station in Hungary 1 above will be utilised along with an additional border station on the Szamos River and a station at the confluence of the Tisza and Bodrog rivers at Tokaj. Detection of contamination at these stations and on the Tisza River near the intake for the East Main Canal serving Debrecen will allow for a timely response by the Debrecen municipality to minimise impacts on its drinking water supply. Other users of canal water can also be notified. Responses include utilisation of reservoir or groundwater for drinking water supply and cessation of irrigation withdrawals. Sampling will be conducted at one additional site just above the confluence of the Tisza and Bodrog rivers at Tokaj.

4.1.3. *Task H3. Romania/Hungary Cross-Border Monitoring*

This task will provide equipment, technical assistance, and training to local authorities to monitor pollutants of concern at four sites in Cris/Körös basin at the Romanian/Hungarian border and to take appropriate actions to minimise impacts on downstream surface water users, especially fish farming near the border and the recharge of bank-fed potable water wells on the Berettyó River arid of shallow wells in rural areas. Sampling will be conducted at 4 sites: one each on the Black, White, and Rapid Cris/Körös and the Barcau/Berettyó River at the border with Romania.

4.1.4. *Task H4. Barcau/Berettyó River Irrigation Protection*

This task will provide equipment, technical assistance, and training to local authorities in order to develop and implement a rapid response plan and communication network in connection with use of the Barcau/Berettyó River for irrigation purposes.

Early detection will allow for the timely suspension of irrigation withdrawals for short periods to minimise the use of polluted surface water during periods of high concentrations of contaminants. The main contaminant of concern is oil from the Suplacu oil complex in Romania. This task will increase the ability of responsible agencies to monitor the Barcau/Berettyó River for increases in levels of oil, petroleum products, and phenols at the Romanian border and will permit accident warnings to be issued to downstream users.

4.1.5. *Task H5. Rapid Körös River Well Protection*

This task will provide equipment, technical assistance, and training to local authorities in order to develop and implement a rapid response plan and communication network to mitigate impacts of pollution of the Rapid Cris/Körös River on subsequent groundwater withdrawals and a nearby fish farm. The main contaminants of concern are heavy metals, phenol, pathogens, etc. from the Oradea municipality and industries.

4.1.6. *Task H6. Black/White/Double Cris/Körös Irrigation Protection*

This task will provide equipment, technical assistance, and training to local authorities in order to develop and implement a rapid response plan and communication network, to mitigate impacts of pollution of the Black, White, Double Cris/Körös River on subsequent irrigation withdrawals. The main contaminants of concern are radioactivity from uranium mining and heavy metals from non-ferrous mining operations in Romania.

4.1.7. *Task H7. Maros/Muresul River Recharge Protection*

This task will be a study of the Maros/Muresul River, with special attention to the "Old Maros" aquifer. to assist local authorities in planning final projects that will reduce pollution of the aquifer. An MLIM station will be established on the Maros/Muresul River at the point where the river becomes the common border between Romania and Hungary.

The Maros/Muresul River alluvium provides a large drinking water source for Hungary and Romania. Approximately 80% of the aquifer is located in Romania, and the remainder is in Hungary. This resource is more attractive than the deep wells in Békés county which are contaminated with naturally occurring arsenic concentrations; however, it would be necessary to ensure that the quality of the water taken from the Maros/Muresul Aquifer is better than that of the Békés wells. Study of the Maros/Muresul aquifer will provide useful information on its quality and quantity, and will help in the development of the necessary steps for its protection and use.

4.2. SLOVAKIA: TWO TASKS

4.2.1. *Task S1. Ukraine/Slovakia Cross-Border Monitoring & Lekarovce Well Protection*

This task will provide equipment technical assistance and training to local authorities to monitor pollutants of concern on the Uh River at the Ukrainian/Slovak border and to take appropriate actions to minimise impacts on downstream water users, particularly the bank-fed wells utilised by the city of Lekarovce.

Responses to detection and early warning from this station would include cessation of well field withdrawals, start-up treatment processes such as the ultraviolet radiation/ozonation process currently being pilot tested, and spill containment measures.

This intervention will increase responsible agencies' ability to monitor the Uh River for periodic high levels of petroleum, phenols, nitrites, organics, PCB's, and radioactivity at the Ukrainian border.

4.2.2. *Task S2. Kosice Municipal Wastewater Treatment Plant Improvements*

The purpose of this task to provide equipment, technical assistance, and training in order to upgrade the Kosice Wastewater Treatment Plant. Improvements to the aeration system, process instrumentation, metering, and controls are anticipated.

Pathogens, phenols, and other pollutants now released by the Kosice Wastewater Treatment Plant to the Hornad River in Slovakia, will be reduced by this intervention, lowering the risk of pathogens and other pollutants reaching the human food chain downstream in both Slovakia and Hungary.

4.3. ROMANIA - THIRTEEN TASKS

4.3.1. *Task R1. Suplacu Petrom/Petrochem(A).-Monitoring of industry by Apele Romane*

This task will provide equipment, technical assistance, and training to local authorities to enhance their ability to monitor and regulate discharges from oil production and refining facilities in Bihor County.

Additional sampling and local analytical capability by Apele Romane will help to ensure compliance with discharge regulations by the oil production and refining operations. Local analysis of samples will promote a rapid response to problems.

Improvement in the ability of responsible agencies to monitor and regulate the release of oil, phenols, and other organics to the Barcau/Berettyó River from the Suplacu oil fields and refinery in Romania will result in a reduction of toxics reaching downstream water users in both Romania and Hungary. Data thus collected will also support accident warning and response interventions in Hungary.

4.3.2. *Task R2. Suplacu Petrom/Petrochem (B): Industrial Phenol Reduction and Runoff Management and Treatment*

This task will provide equipment, technical assistance, and training to officials at Suplacu Petrom/Petrochem to improve both wastewater treatment capabilities as well as containment of contaminated runoff at the oil production and refining facilities.

Two activities are proposed. The first will provide additional wastewater treatment capabilities and more efficient operating and waste minimisation practices for both the production and refining operations. The second will involve collection and treatment of runoff from the oil production and refining operations. Both activities will help to minimise contamination of the river with phenol and petroleum products. In addition to short-term improvements, a passive treatment system will provide a level of protection even after the oil facilities are no longer in use.

Oil and phenols now released to the Barcau/Berettyó River from process discharges, spills, and storm water runoff from Suplacu oil fields and refinery will be reduced by this intervention.

4.3.3. *Task R3. Oradea Monitoring Support: Environmental Protection Agency*
Task R4. Oradea Monitoring Support: Apele Romane

These two tasks will provide equipment, technical assistance, and training to local authorities to enhance their ability to identify surface water quality problems throughout the Cris basin.

Additional sampling and analytical capability will help the Environmental Protection Agency to identify surface water quality problems and will help Apele Romane to ensure compliance with regulators by municipalities and industries. Selection of major equipment (e.g., gas chromatograph [GC]) or atomic absorption spectrophotometer [AA]) to be provided in Oradea to the Environmental Protection Agency and to Apele Romane will be designed to avoid duplication and to foster co-operation between the two laboratories.

These agencies have responsibility for the Barcau/Berettyó River and the Black, White and Rapid Cris/Körös Rivers. Improvement in their abilities to monitor and regulate the release of pathogens and toxic substances including heavy metals, oil, phenols and other organics to the various rivers in the basin should result in a reduction of these pollutants from reaching downstream water users in both Romania and Hungary.

4.3.4. *Task R5. Oradea Municipal Wastewater Treatment Plant (A) - Sludge Digestion, Drying and Management*
Task R6 Oradea Municipal Wastewater Treatment Plant (B): Lagoon System Redesign and Modification
Task R 7. Oradea Municipal Wastewater Treatment Plant (C): Oil and Grease
Task R8, Oradea Municipal Wastewater Treatment Plant (D):O&M and Technical Assistance

These four tasks will provide technical assistance and equipment in order to upgrade the current Oradea Wastewater Treatment Plant and to provide technical assistance and training in improved operation and maintenance. An upgraded, well-operated and well maintained Oradea Wastewater Treatment Plant will reduce contamination of the Rapid Cris/Körös River by pathogens, solids, oxygen demanding compounds, and, to a lesser extent heavy metals, and toxic organics. This activity will also promote institutional strengthening of a system for user fees and fines for industrial contributors, transfer of pre-treatment technology to industrial contributors, especially regarding the control of heavy metals received at the plant, and development of a sludge management plan.

4.3.5. *Task R9. Oradea Pigment industry: Chromium and Zinc Removal/Waste Minimisation*
Task R10. Oradea Pharmaceutical Industry: Phenol Removal/Waste Minimisation

Task R11. Oradea Metalworking Firms: Waste Minimisation Technical Assistance

The purpose of these tasks is provide equipment, technical assistance, and training to industry officials in industrial waste minimisation as well as improved operation and maintenance of industrial pre-treatment facilities.

The effluent from the pigment factory typically contains less than the regulated concentration of hexavalent chrome; however, the treatment process is unreliable and expensive. Treatment process control equipment will increase reliability and decrease cost. Regulations requiring that the effluent from the pigment factory contain less than 0.1 mg/L of zinc are regularly violated by up to 10 times. Waste minimisation and wastewater treatment will be used to reduce zinc contamination in the river.

Heavy metals and phenols now released to the Oradea Wastewater Treatment Plant and the Rapid Cris/Körös River from industries in Oradea will be reduced by this intervention. Reduction of levels of heavy metals discharged by the factories will lower risks of these substances reaching the human population downstream in Romania and Hungary.

4.3.6. *Task R12. Uranium Mining: Industrial Tailings Management and Monitoring*

These tasks will provide equipment, technical assistance, and training designed to remove uranium and non-ferrous contamination from mine drainage or ore processing waters and to cap or treat runoff from abandoned mining tailings piles.

Drainage from dewatering processes in uranium mining and runoff from a separate area of uranium tailings abandoned for decades, enters a tributary of the Black Cris/Körös. An overloaded sediment pond partially removes radioactive sediment from the mine drainage water. Treatment to remove uranium from mine drainage water and capping and/or treatment of runoff from abandoned tailings piles will reduce radiological contamination of the Black Cris/Körös.

4.3.7. *Task R13. Non-Ferrous Metals Mining: Trace Metals Removal/Pollution Prevention*

This task will provide equipment, technical assistance, and training to remove heavy metals and cyanide from non-ferrous mine drainage and ore processing wastewater from a tailings pond, currently discharged to a tributary of the Black Cris/Körös River.

Heavy metals released into the Black Cris/Körös River from ongoing mining activities in Romania will be reduced by this intervention, thereby lowering risk to public health downstream.

5. Conclusions

This project will provide various water inspectorates in Slovakia, Hungary. and Romania with valuable tools for monitoring baseline water quality and for assessing positive or negative changes in water quality over time. From the data collected by the inspectorates, the national governments will know to a large degree how much pollution is crossing to or from their inspectorates and will also have warning of impending spikes of pollution which may threaten downstream water users. They should then be able to give early warning to those to curtail use until the contamination passes. Analytical equipment to be provided will help inspectorates to reliably analyse water samples in the field or in the laboratory.

Municipal and industrial wastewater treatment plant improvements will improve the quality of water downstream and reduce threats to human health and to the food chain in the tributaries. Training will ensure that the capital investment in equipment will be matched by competent knowledge of the operation and maintenance of system equipment.

6. Recommendation

It would be most helpful to institute national regulations to waive the import duties on equipment funded or donated for use on projects which have health and environmental benefits such as this project. Significant time and effort which is expended on resolving issues of import duties/taxes could then more productively be used towards implementing the project.

Acknowledgements

Portions of this paper were adapted, with permission, from the project scope written by Dr. William Hogrewe, formerly with USAID, and Carl Mitchell USAID, Washington.

References

1. WTV, Monitoring, Laboratory Analysis and Information Management for the Danube River Basin *Final Report* **Volume IV**, Summary, 21 February 1994,
2. Ibid., **Volume I**, Priority Needs to Strengthen the Danube Basin Monitoring Network; February 1994.
3. Delft Hydraulics, Environmental Programme for the Danube River Basin, Accident Emergency Warning System, Set-up of system, *Final Report*, September 1994

The following is a summary of the various project tasks by country. Those marked with an asterisk (*) specifically include the establishment of MLIM and AEWS stations and assistance to inspectorates in increasing their analytical capabilities.

ACTIVITIES IN HUNGARY

H1*	Slovakia/Hungary Cross-border Monitoring
H2*	East Main Canal/Debrecen Drinking Water Protection
H3*	Romania/Hungary Cross-border Monitoring
H4*	Barcau/Berettyó River Irrigation Protection

H5	Rapid Cris/Körös River Well Protection
H6	Black/White/Double Cris/Körös Irrigation Protection.
H7	Maros/Muresul River Recharge Protection

ACTIVITIES IN SLOVAKIA

S1*	Ukraine/Slovakia Cross-Border Monitoring & Lekarovce Well Protection
S2	Kosice Municipal Wastewater Treatment Plant Improvements

ACTIVITIES IN ROMANIA

R1*	Suplacu Petrom/Petrochem (A): Monitoring of Industry by Apele Romane
R2	Suplacu Petrom/Petrochern (B): Industrial Phenol Reduction and Runoff Management and Treatment
R3/R4*	Oradea Monitoring Support: Environmental Protection Agency and Apele Romane
R5/R6/R7/R8	Oradea Municipal Wastewater Treatment Plant Improvements
R9	Oradea Pigment Factory: Chromium and Zinc Removal/Waste Minimisation
R10	Oradea Pharmaceutical Factory: Phenol Removal/Waste Mininisation
R11	Oradea Metalworking Firms: Waste Minimisation Technical Assistance
R12	Uranium Mining-, Industrial Tailings Management and Monitoring
R13	Non-Ferrous Metals Mining: Trace Metals Removal/Pollution Prevention

AUSTRIAN - GERMAN DATA SHARING

SCHWAIGER K.
Federal Ministry of Agriculture and Forestry
Department of Water Management
Marxergasse 2
A-1030 VIENNA

KEY WORDS / ABSTRACT: Inn River Basin / Austrian-German / Water Quality Data-Sharing / Danube River

A brief outline of the core parts of the legal basis for as well as the range of practical experiences arising from water quality monitoring is followed by a description of bilateral Austrian-German projects such as conservation of wetlands, ground water modelling, among others. The focus is on the sharing of water quality data, the benefits of tailor -made monitoring and quality assurance programmes.

1. Geographic position

Germany and Austria share a border more than 650 kilometres long and a number of major transboundary rivers, all of them draining into the Danube Basin. About 150 kilometres of the border follow, in part, the course of the Rivers Inn, Salzach and Danube.

All transboundary rivers in this region flow northwards from the Austrian Alps and discharge into Bavaria. Considerably enlarged by the discharge from the Bavarian territory, those waters reenter Austria via the Danube in Jochenstein. Germany and Austria are therefore simultaneously upstream as well as downstream countries from each other.

2. Legal basis for co-operation in water management

Due to this very specific situation and the resulting mutual interdependence, technical and legal co-operation in water management affairs on the state level has a long tradition.

One agreement still in force, for example, dates from 1923 and concerns the transfer of water from River Lech, a tributary to the Danube, into the Main and from there via Rhine to the North Sea.

I. Lyons Murphy (ed.), Protecting Danube River Basin Resources, 165–171.

The topics dominating bilateral co-operation immediately after the second world war were agreements concerning the generation of hydroelectricity in transboundary rivers. Several such treaties were put into force in the fifties. In the early seventies the project of the Rhine - Main - Danube water way dominated bilateral co-operation.

Due to growing pressure on freshwater resources, especially on freshwater quality and due to growing environmental concern, both countries convened to add an institutional framework to already existing excellent neighbourly contacts. This treaty concerning bilateral water management affairs in a very comprehensive way was signed in Regensburg in 1987 (therefore also called treaty of Regensburg) by the federal Republic of Germany, the Federal Republic of Austria and the European Community. This treaty came into force 1991.

2.1 TREATY OF REGENSBURG:

In this treaty, both neighbouring countries and the European Union agreed among others items to:

- co-operate intensively in the field of water management,
- exchange all relevant information,
- to consult each other on all projects which might result in major damage to the neighbouring country,
- monitor water quality in transboundary rivers,
- harmonise existing or jointly devise new emergency warning systems,
- install a permanent transboundary river commission,
- settle disputes within an arbitration tribunal as well as those concerning detailed statutes for this commission.

The transboundary river commission was installed in 1991. Due to the federal structure of both neighbouring countries members from both state governments as well as from Bavaria, Tyrol, Salzburg and Upper Austria and from the European Community are represented.

Within the transboundary river commission and its subgroups all fields of work listed in the treaty have been covered in an extremely constructive, friendly and successful way. Among others, a monitoring network covering all transboundary rivers has been installed, a comprehensive monitoring programme including joint sampling has been realised, data on freshwater quality have been exchanged regularly. existing agreements have been reviewed, important studies have been initiated (e.g. on deep transboundary groundwater aquifers, preventive measures to stop bed corrosion in the River Salzach...) and problems arising from waste water management have been treated.

The settling of existing minor problems within the transboundary commission has become easier due to existing similar approaches to water management problems (abatement of emissions at the sources based on BAT combined with control of immissions...) in both countries.

2.2. IMPACT OF JOINING EUROPEAN UNION

Despite optimistic expectations, no real new dimensions and perspectives were introduced into the bilateral co-operation when Austria joined the European Community, as existing water legislation of the Community has proved to be too inconsistent (at present, a revision of existing water directives as well as a new framework directive is discussed at Community level) and less specific than the water legislation of both neighbouring countries.

Referring to the list of activities agreed on within the treaty, the sharing of water quality data is only one of several important items to ensure good neighbourly relations within the broad fields of bilateral water management. Nevertheless, details will be presented only as regards the main topic in order to stick to the initial subject attributed to the author.

3. Fresh-water quality monitoring

3.1. MONITORING NETWORK

All transboundary rivers, irrespective of their importance, are subject to bilateral freshwater quality monitoring. Appropriate sites in the proximity of the borderline have been selected jointly. Where the borderline follows the course of the rivers, representative sites have been agreed upon. Important details of the selected sites such as name of station, name of river, size of catchment area at sampling site (km^2), average annual discharge (m^3/s), the state in charge of monitoring, further national and international monitoring obligations (e.g. due to 77/795/EEC-Exchange of information, Bucharest Declaration, Transnational Monitoring Network for the Danube basin) are listed in Table 1.

As can be seen from Table 1, monitoring efforts are shared equally by both countries. Nearly all sites are also part, respectively, of the national Austrian and Bavarian quality network. Some sites (e.g. Jochenstein, Oberndorf - Laufen, Kufstein) are also part of other international monitoring activities.

Jochenstein on the River Danube is of key importance to national and international monitoring programs. In Jochenstein an average annual discharge of 1,420 m^3/s from a catchment area of more than 77 000 km^2 enters Austria via the Danube. About 16,000 km^2 (20%) of the 77,000 km^2 are in Austrian territory (Tyrol, Salzburg, Upper Austria), about 77% German and a small share are in Swiss territory. Although Austria has a share of only about 20% of the catchment area at Jochenstein, more than 45% of those 1,420 m^3/s are of Austrian origin due to abundant precipitation in the Austrian part.

3.2. MONITORING PROGRAMME - SUBSTANCES AND FREQUENCIES

Against with trends observed internationally at that time, a pragmatic approach with lists of substances tailor-made to the specific regional and local situation of every site has been agreed on. The details of this programme are listened in Table 2.

TABLE 1: Monitoring network, site characteristics

site	river	catchment[1] area (km^2)	discharge[1] (m^3/s)	monitoring responsibility	other monitoring obligations
Füssen	Lech	1417	(58)	G	BM, AM
Ehrwald	Loisach	88	3	A	-
Mittenwald	Isar	286	9	G	BM, AM
Erl/Kufst.	Inn	9905	306	A	AM, BM, EU
Kössen	Großache	821	30	A	AM, BM
Unken	Saalach	857	33	A	AM, BM
Salzburg	Saalach	1139	44	A	AM, BM
Laufen	Salzach	6113	239	G	BM, EU, AM
Eschelbach	Inn	13354	369	G	BM
Braunau	Inn	22841	(643)	G	BM, AM
Passau	Inn	26049	734	G	BM, AM
Obernzell	Donau		1420	A	-
Felsenhütt	Donau		1420	A	-
Jochenstein	Donau	77086	1420	A-G	AM,BM,B, TNMN

Abbreviations used:

A	Austria
G	Germany
AM	Austrian national quality monitoring network
BM	Bavarian quality monitoring network
B	Bucharest Declaration.
TNMN	Transnational Monitoring Network
EU	77/795/EEC-Exchange of Information about Surface Fresh Waters
[1])	Hydrographisches Jahrbuch v. Österreich 1992, Hydrographisches Zentralbüro, Wien 1995 Deutsches Gewässerkundliches Jahrbuch, Donaugebiet, 1990, Bayrisches Landesamt für Wasserwirtschaft, München

A set of descriptive basic parameters, nutriments included, are monitored at least monthly at every site. Sets of specific parameters (e.g. chlorinated hydrocarbons, heavy metals, radioactivity) are also monitored monthly, but only on sites downstream of important industrial or communal agglomerations. Saprobity is determined once a year on all sites, bacteriological parameters are monitored 4 times per year on all sites.

3.3. IMPLEMENTATION OF THE MONITORING PROGRAMME

All monitoring activities were initially assigned to governmental institutions in both countries. At the very beginning of the monitoring programme all sites were visited jointly in order to agree on the exact location for taking samples. The samples taken on that occasion were analysed by each side. The results obtained by each side were compared and proved to be identical. As a consequence, neither side believed it was necessary to continue with joint sampling. With the exception of Jochenstein, no joint sampling has taken place since that very first round of monitoring. By avoiding regular joint sampling, considerable savings could be realised, as joint sampling along a borderline of 650 kilometres requires close co-ordination and a lot of additional time.

TABLE 2 - Monitoring programme

site	river	resp.	basic	DOC	chlor	metals	bact.	rad.	biol.
Füssen	Lech	G	x	-	-	-	x	x	x
Ehrwald	Loisach	A	x	-	-	-	x	-	x
Mittenwald	Isar	G	x	-	-	-	x	-	x
Kufstein	Inn	A	x	x	-	x	x	x	x
Kössen	Großache	A	x	-	-	-	x	-	x
Unken	Saalach	A	x	-	-	-	x	-	x
Salzburg	Saalach	A	x	-	-	-	x	-	x
Laufen	Salzach	G	x	x	x	x	x	x	x
Eschelbach	Inn	G	x	x	-	-	x	-	x
Braunau	Inn	G	x	x	-	-	x	-	x
Passau	Inn	G	x	x	-	x	x	x	x
Obernzell	Donau	A	x	x	x	x	-	x	x
Felsenhütt	Donau	A	x	x	x	x	x	x	x
Jochenstein	Donau	G,A	x	x	x	x	x	x	x

Abbreviations:

resp.: state responsible for monitoring at site (A ... Austria, G ... Germany) basic: descriptive basic parameters (temperature, pH., conductivity, dissolved oxygen, oxygen saturation rate, hardness, calcium, magnesium, alkalinity, biochemical oxygen demand, chemical oxygen demand (KMNO), orthophosphate, total phosphorous, ammonia, nitrate, chlorides, sulphates, total suspended solids,)

DOC.: dissolved organic carbon

chlor.: chlorinated hydrocarbons: (chloroform, tetrachlorethene, trichlorethene, tetrachlormethane, 1,1,1 - trichlorethane; AOX)

metals: heavy metals (Zinc, total iron, total manganese, lead, cadmium, copper, total chromium, nickel, arsenic, mercury)

bact.: bacteriological parameters: (total coliforms, faecal coliforms, faecal streptococci, salmonella, number of colonies);

bio.: biology (saprobity)

rad.: radioactivity: (total alpha, total beta, tritium, gamma, strontium 90)

A report with the monitoring results of both sides is collated by the two governmental organisations in charge of monitoring activities alternately; one year by the "Bayrisches Landesamt für Wasserwirtschaft" and the subsequent year by "Bundesamt für Wasserwirtschaft - Wien". The necessary data is provided by the partner organisation on a diskette in an agreed format. The set of data for the whole year is assessed jointly by experts from both sides. In the yearly meeting of the transboundary commission the report with all findings and all monitoring data is adopted and exchanged officially.

3.4. EXPERIENCES WITH DATA SHARING

As shown in Table 1, nearly all sites are also part of the Austrian national network for monitoring river water quality. In Austria a national network consisting of 245 sites was set up in the year when bilateral monitoring of transboundary rivers started. This national network is organised by the Federal Ministry of Agriculture and Forestry in Vienna in close co-operation with Austria's 9 provinces. Sampling and analysing is nearly entirely performed by private top laboratories, which are awarded their contracts by public invitations to tender. As the capacities of private laboratories have been used at no cost, extremely low costs for their services were possible.

In the first year national monitoring and bilateral monitoring of transboundary rivers were performed independently in Austria. This duplication of efforts was stopped as both systems forwarded identical results and as permanent quality assurance programmes (based on interlaboratory calibration tests, spiked samples and audits within the laboratories) proved to be at the same high level of analytic performance.

As Austria's "Bundesamt für Wasserwirtschaft - Wien" is located 300 to 600 kilometres from the common border, sampling and analysis of transboundary rivers is now entirely performed within the national network. The responsibility for collating data and the drafting of reports still remains within the Bundesamt.

To avoid slightly different databases in both countries, all existing relevant information is exchanged. As Bavaria is monitoring biweekly, all 26 results of the relevant year and not only the monthly results are exchanged.

4. Conclusions

Monitoring of transboundary rivers can be an extremely time consuming and expensive task. Monitoring programs with lists of substances tailor-made to the very specific local and regional situation can save many fruitless efforts. Thus, costs arising from monitoring are accepted more willingly. Abandonment of joint sampling can improve flexibility and save costs, but requires high quality performance as well as mutual confidence on the part of all sides involved in monitoring. Quality assurance schemes based on national as well as on international interlaboratory comparison tests, rounds of spiked samples and internal and external audits of laboratories are a prerequisite to produce reliable and fully comparable data. The sharing of data between Austria and Germany is one example of how the manifold problems related to monitoring have been solved in a good neighbourly spirit satisfying both sides.

References

1. BGB1. Nr. 17/1991, Vertrag zwischen der Republik Österreich einerseits und der Bundesrepublik Deutschland und der Europäischen Wirtschaftsgemeinschaft andererseits über die wasserwirtschaftliche Zusammenarbeit im Einzugsgebiet der Donau (Austrian OJ Nr/1991, *Treaty of Regensburg*)
2. Sachverständigen-Arbeitsgruppe Gewässerschutz der Ständigen Gewässer-kommission, Regensburger Vertrag, Untersuchungsergebnisse 1992, 1993, 1994 (*Working group on water protection of the Permanent Transboundary River Commission, Treaty of Regensburg*, Monitoring results 1992, 1993, 1994).
3. Wasserwirtschaftskataster/Umweltbundesamt (1995), Wassergüte in Österreich - Jahresbericht 1994, Bundesministerium für Land- und Forstwirtschaft, Wien (*Federal Water Management Register/Federal Environmental Agency 1995, - Freshwater Quality in Austria - annual report 1994*, Federal Ministry of Agriculture and Forestry, Vienna).

DISCUSSION FOLLOW-UP

With respect to the core set of sampling at specific sites, participants asked if parameters are reviewed; the reply from Mr. Schwaiger was that they were reviewed on a yearly basis. Had it been possible to reduce the number of parameters for which you monitor? Schwaiger responded with an example. Emissions in the paper industry had changed therefore so that it was possible to delete some formerly essential parameters; it was necessary, however, to keep a very watchful eye.

When asked about NGO or green organisation issues, Schwaiger replied that the Alpine rivers are very clean even with the high population density and other factors. Data on monitoring results are publicly available.

If rivers are clean why is bacterial content measured? Schwaiger replied that they had to monitor waste water treatment plants and that it doesn't cost much. This is a standard practice in Germany as well (Liebscher). In response to other questions Schwaiger said that the waste water treatment plants do not disinfect but only remove nutrients.

If relationships are so harmonious and you have clean rivers why do you have a treaty? It helps to have a broader level of exchange. We had already had some 25 agreements and the one we have name provides a general framework, institutional background and building, according to Liebscher. The questioner suggested they were putting into an institutional framework a number of agreements. Liebscher said that such agreements are necessary to establish the operation because the finance ministry always asks why do it, with a treaty it means a decision has been made to "do it."

A participant asked if monitors had detected a strong input of bacterial contamination other than from point sources? Most are from point sources. Is there control at the site and do you expect changes? It was suggested to Mr Schwaiger that urban water run-off, with bacterial contamination is a big worry in the Danube and if waste water treatment isn't successful, it can be a real problem. Mr. Schwaiger agreed that it was one of the problems.

He also said that the Austrian-German Treaty has been signed much earlier but nevertheless it covered the substance of the basic Danube Convention and he saw no need to rewrite it.

DATA SHARING BETWEEN THE SLOVAK AND CZECH REPUBLIC IN THE MORAVA RIVER BASIN

PETER RONČÁK
MARCELA FABIANOVÁ
Slovak Hydrometeorological Institute
Bratislava, the Slovak Republic

JOZEF DROZD
Morava River Basin Authority
Brno, the Czech Republic

KEY WORDS / ABSTRACT: Morava River Basin / Dyje River Basin / Czech Republic / Slovak Republic / water quality monitoring / water quality data sharing / Danube River Basin

This paper provides an overview of activities related to surface water quality, mainly its monitoring within the Morava River Basin with special attention to the exchange of data and information between the Slovak Republic and the Czech Republic. It includes basic information on the Morava River Basin, monitoring network and international activities related with surface water quality.

1. Background information

The River Morava is one of the most important tributaries of the Danube River. The total area of the Morava River Basin is 26,658 km^2 of which approximately 8.4% belongs to the Slovak territory and about 84% to the Czech territory. The Morava River and its biggest tributary, the Dyje, create one hydrological unit and their basins spread over three states - the Czech Republic, Austria and the Slovak Republic (Figure 1).

The Dyje River rises in the Czech Republic; as the Moravian Dyje it flows into Austria. After its confluence with the Austrian Dyje it flows again to the Czech Republic where it flows into the Morava River. The Dyje River forms a border at its mid-point; before its confluence with the Morava River it is 40 km long.

The Morava River rises in the Czech Republic and flows into the Slovak Republic. Its border length is approximately 110 km The length of the border part of the river is approximately 30 km between the Slovak and the Czech Republic.

I. Lyons Murphy (ed.), Protecting Danube River Basin Resources, 173–181.

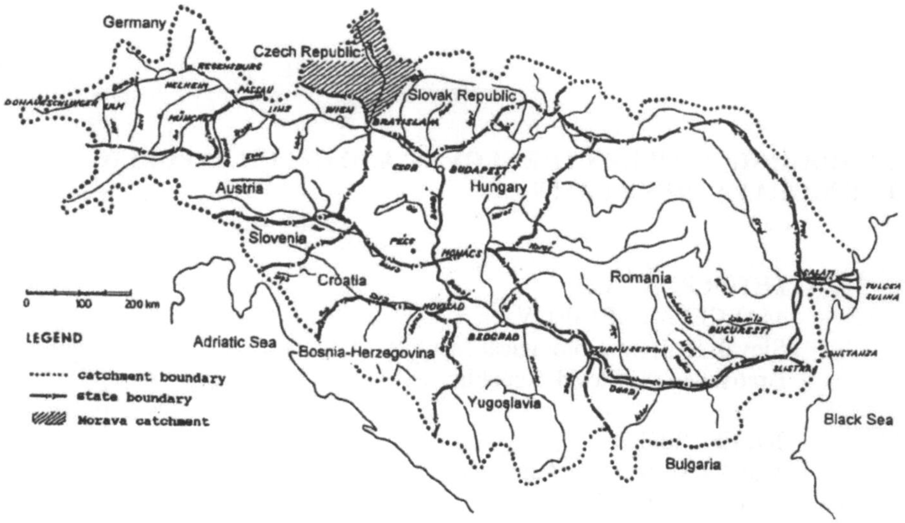

Figure 1. The Danube River Basin

The downstream part of the Morava River creates a border between the Slovak Republic and Austria approximately 70 km in length up until its confluence with the Danube River. The total length of the Morava River is 352 km.

From this short description it is clear that monitoring and quality control of surface waters is not and cannot be the problem of only one state, but it is at least bilateral issue. Transfer of information and data exchange between involved partners play a very essential role.

2. Monitoring activities

2.1. DEVELOPMENT

Due to the common historical development of the Slovak Republic and the Czech Republic, systems of environmental quality monitoring are very similar. Surface water quality has been monitored since 1963 in both the Slovak and the Czech Republic. The results of water quality monitoring are presented in yearbooks of surface water quality. They were published for both the Slovak and the Czech Republic together from 1963 to 1967. Currently, the responsible publishers of water quality yearbooks are the Slovak Hydrometeorological Institute (SHMI) and the Czech Hydrometeorological Institute (CHMI).

The following institutions are involved in surface water quality monitoring activities: the Ministry of Environment of the Slovak Republic (SHMI) and the Ministry of Environment of the Czech Republic (CHMI).

These two ministries administer monitoring networks, and are responsible for all activities associated with monitoring such as sampling, analysis, data management, and the presentation and evaluation of information on surface water quality. The co-operation between both HMIs is not official, e.g. free exchange of water quality yearbooks is not customary. Regarding data on water quantity, they are available for relevant hydrological stations in the Czech part of the Morava Basin.

The Danube River Basin Authority and the Morava River Basin Authority constitute bodies which administer river basins have responsibilities to monitor the water quantity and quality of surface streams. Currently these bodies have been nominated by HMIs for sampling and laboratory analyses.

2.2. MONITORING NETWORK

In general water quality is monitored on two levels:

- national: state or main water quality sampling sites
- regional: water quality sampling sites on water streams and in water reservoirs which are directly under the administration of the basin authority and serve for internal decision making.

Other information deals only with national monitoring network, what is relevant from the point of view of data and information exchange.

SHMI and CHMI have the most important role from the practical point of view on data exchange. As was mentioned before, these organisations are responsible for the administration of monitoring networks. In practice it means that requirements on the monitoring network are re-evaluated and steps taken to improve the usefulness of sampling sites carried out by HMI every year. Plan of water quality monitoring is adjusted every year with regard to water management needs capability and financial aspects are also taken into consideration.

Water quality monitoring network is oriented on the most critical parts of streams which are influenced by discharged pollution. The network allows:

- to identify flow of pollution;
- to evaluate long-term trends in water quality development;
- to give preliminary information to assess water quality for its further uses;
- hazard assessment.

In the Slovak part of the Morava River Basin water quality was monitored at 17 basic sampling sites in 1994. Four of them are located directly on the Morava River and the rest on main tributaries.

In the Czech part of the Morava and Dyje River Basin water quality is monitored at 62 sampling sites of the state monitoring network, 11 of which are situated directly on the Morava River and the Dyje River, 59 regional and 85 sites on reservoirs (Figure 2).

Sampling and chemical analysis of samples are carried out by the river basin authorities. The results are provided to HMIs which are responsible for additional management of the data (validation, storing, reporting and interpretation). In SHMI the data on water quality are stored, processed and evaluated in the database system

MAGIC by the Water Quality Department. In the CHMI database system ORACLE is employed.

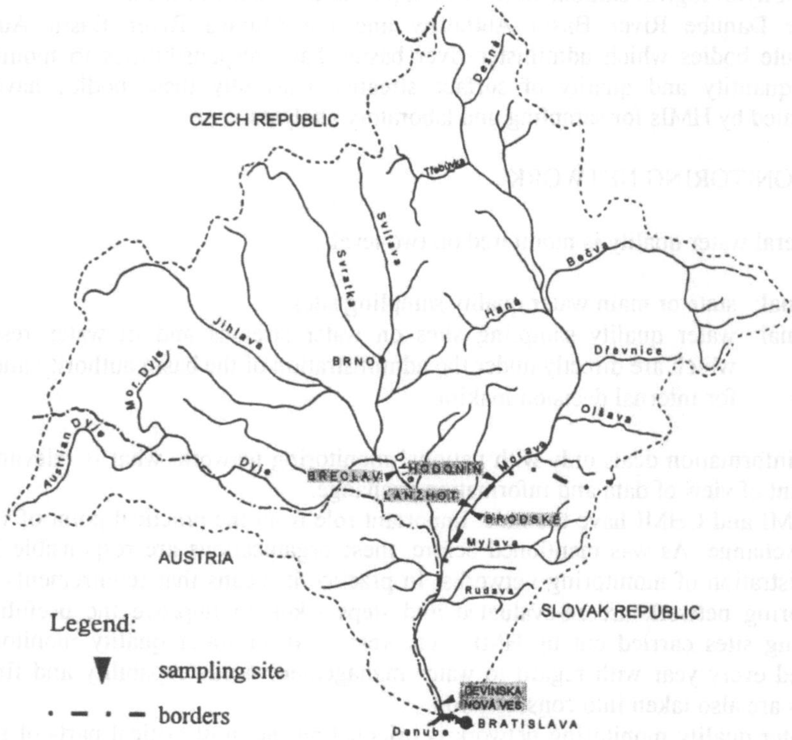

Figure 2. Morava River Basin and selected sampling sites

2.3. ASSESSMENT OF WATER QUALITY

The water quality of surface streams is evaluated according to the former Czechoslovak standard ČSN 75 7221 classification of surface water quality, which was adopted in the Czech Republic in 1989 and in the Slovak Republic in 1990.

Monitored water quality determinands are divided into 6 groups (table 1) and water quality is classified into 5 classes according to these parameters.

TABLE 1. Monitored determinands according to STN 75 7221

	Group of determinands	Determinands
A	Oxygen regime	dissolved oxygen, BOD_5, $CO_{DMN,Cr}$ (organic carbon, sulphate, sulphide)
B	Basic chemical and physical determinands	pH, temperature of water, soluble compounds (conductivity), non-soluble substances, $N\text{-}NH_4$, $N\text{-}NO_3$, P_{total} (total iron, manganese, organic nitrogen)
C	Supplementary chemical determinants	calcium, magnesium, chlorides, anionic surfactant, nonpolar extractable compounds (NEL), organically bounded chlorine (absorbency, fluorides, phenols -volating with water steam, total cyanide, active chlorine, extractable organically bounded chlorine)
D	Heavy metals	mercury cadmium, arsenic, lead (copper, total chromium, chromium VI, cobalt, nickel, zinc, vanadium, silver)
E	Biological and microbiological determinants	saprobity index, coliform bacteria of faecal coliform bacteria (psychrophile bacteria, enterococcuses)
F	Radioactivity	total volume alpha activity, total beta activity (radium, uranium, tritium)

Sets of monitored determinands are provided by HMIs with respect to monitoring needs. It means that not all determinands are monitored in all sampling locations (e.g. groups D, F), but where and when necessary. There are also determinands (e.g. pesticides, oil substances etc.) which do not belong to the identified standard, but to other water-quality related standards or prescriptions. They are monitored at relevant sampling sites e.g. in areas with intensive agricultural activities.

Frequency of sampling is for most of the determinands usually 12 times per year; for some of them it is 4 or 6 times per year.

Data on heavy metals were not sufficient to classify water quality in the group of determinands "D" in 1993-94. Radioactivity is not monitored.

The Morava River belongs to the group of rivers which are strongly impacted by pollution (table 2). It is due to pollution sources located in the Slovak Republic but also because of pollution coming from the Czech side.

In order to improve and maintain surface water quality within the Morava Basin an improvement in the exchange of information between the Slovak Republic and the Czech Republic is necessary. Such information would help to identify the amount, type and source of pollution.

The last surface water quality sampling site in the Czech part of Morava is in 81.0 river km and the first sampling site in Slovak side of Morava is Morava-Brodské in 79.0 river km.

TABLE 2. Water quality classification of Morava River in 1993-1994

Stream - profile	type of profile since year*	river km	group of determinants quality class	the most adverse determinants	Remark
Morava - Devínska Nová Ves	basic 1967	1.50	A - IV	COD	confluence with the Danube River
			B - V	$N\text{-}NO_2$	
			C - IV	phenols volating with water steam	
			E - V	Psychrophyle bacteria	
Morava-Brodské	basic 1967	79.0	A - III	BOD, COD	the first quality profile in Slovak part of Morava
			B - V	$N - NO_2$	
			C - IV	phenols	
			E - V	psychrophile bacteria	
Morava Lanhot	basic 1967	81.0	A - III	BOD, COD	the last sampling site in Czech part of Morava
			B - V	$N\text{-}NO_2$	
			C - IV	NEL	
			D - V	Hg	
			E - III	coli	
Morava Hodonín	basic 1967	114.0	A - IV	COD	
			B - V	$N\text{-}NO_2$	
			C - II	SO_4	
			E - III	coli	
Dyje Břeclav	basic 1967	26.8	A - IV	BOD, COD	the last sampling site before the confluence with Morava
			B - V	$N\text{-}NO_2$	
			C - III	SO_3	
			E - IV	coli	

* Data for 1986 are not available for the entire Slovak part of the Morava Basin.

2.4. INTERNATIONAL AGREEMENTS

Up to now general co-operation on transboundary rivers issues has been based on bilateral governmental agreements on water management of transboundary rivers. The Slovak Republic has agreements with Austria (1970), Hungary, Ukraine (1996) and Poland (1958). On the basis of these agreements two commissions were established for each country: 1) a border commission and 2) a commission for transboundary rivers. Following further agreements, the commissions have created subcommissions or working groups in order to carry out their activities.

Because the Morava River provides a border between Slovakia and Austria and also because its confluence with the Danube River is close to the Slovak-Austrian border, co-operation on surface water monitoring between Slovakia and Austria is also important to the Morava River Basin.

Regarding surface water quality there is agreement within the Slovak-Austrian commission to co-operate on transboundary river activities including:

- common sampling in reasonable sites
- evaluation, reporting and exchange of annual reports.

The guidelines for early warning systems are also issued, to guide the procedure for water pollution accidents. The same arrangement has also been made with Hungary.

A very similar situation exists in the Czech Republic which has bilateral agreements with Austria, Germany and Poland. An agreement between the Czech and Slovak Republic on the Morava River Basin is still in preparation. In some respects, this shortcoming has been compensated by the activities in multilateral Programmes like Environmental Programme for the Danube River Basin, as well as by activities under Danube Convention on a Sustainable Use and Development of the Danube River Basin, among others.

3. Conclusions

Some conclusions significant to the improved management of important Danube tributaries are suggested by this review:

- Inasmuch as the Morava River forms the border between the Slovak Republic and the Czech Republic for approximately 40 km, exchange of information and close co-operation on water quality issues are essential for both countries.
- Up to now, water quality data exchange between both the Slovak and the Czech Republic is based mainly on unofficial agreements between SHMI and CHMI representatives.
- There are no common sampling activities which permit comparison of the results of measurements and assure their reliability.
- Sampling on only one side of the river is not sufficient to identify the amount and type of pollution which passes from one country to another.

- Co-operation is lacking between river basin authorities, thus sampling and analysis methods cannot be correctly compared.
- Bilateral agreement between the Slovak and the Czech republic is under preparation.
- Comparability of data on water quality is provided since each uses the same basis for water quality standards in the Czech and the Slovak Republic.

4. Recommendations

Based on these conclusions the following recommendations can be made:
On the regional level

- There needs to be co-operation between the Morava River Basin Authority and the Danube River Basin Authority as organisations which will carry out sampling and laboratory analysis.
- Agreement should be reached on common sampling sites in the boundary reach of the Morava .

On the national level

- An international agreement should be reached on the improved management of water quality in transboundary rivers between the Slovak and the Czech Republic and commissions to implement it created.
- There needs to be a continuous exchange of information on such matters as changes in water quality standards, development of surface water quality policies, and other significant matters.
- Compatible quality assessment and control systems for surface water quality monitoring systems need to be adopted in both SHMI and CHMI.
- Common or similar baseline functions need to be adopted in the development and implementation of standards, their enforcement, and other matters in order to co-ordinate all activities related to water quality issues and monitoring.

References

1. Yearbook of Surface Water Quality in Slovakia in years 1993-1994, *SHMI*, Bratislava 1995
2. National Review Slovakia, *The Ministry of Environment of the Slovak Republic*, Bratislava 1994
3. National Review of Morava Basin in the Czech Republic, *The Morava River Basin Authority*, Brno 1993

DISCUSSION FOLLOW-UP

When asked if the data had been updated for publication Ms. Fabianová said water quality is measured every 2 years and she would try to provide an up-to-date publication.

One questioner pointed out that there are nine "hot spots" in the Czech Republic and zero in Slovakia and wondered if he was right in thinking that this will change? (Very likely.)

Josef Drozd showed a map of the Morava River Basin with water quality sampling points that monitoring also involved water quality in 24 reservoirs, with 14 are used for drinking water resources. He believed that there was a difference between transboundary, bilateral agreements. A waste treatment plant has been built in the biggest hot spot in Olomouc. Plants are being built but some areas are quite industrial with a lot of pollution. The government is doing something but lacks the means to build waste water plants.

With regard to the issue of not seeing the point of data exchange if there is no action plan in place, it was pointed out by Fabianová that there is a need to exchange information not necessarily data and then the need for co-operation, i.e. in the case of accidents

Schwaiger suggested that in Austria, providing for exchange of data and information might be the first step to trigger additional activities.

DATA SHARING ACROSS NATIONAL BOUNDARIES IN THE UPPER TISZA RIVER BASIN

F. LÁSZLÓ
Water Resources Research Centre (VITUKI)
H-1453 Budapest, P.O.B. 27, Hungary

KEYWORDS / ABSTRACT: Upper Tisza River Basin / Tisza River Basin / Water Quality Monitoring / Data-Sharing

A description of the Upper Tisza Basin which is shared by Ukraine, Romania, Slovakia and Hungary is followed by a review of present data sharing practices in the Upper Tisza river basin including the exchange of hydrological and hydrometeorological data and water quality data. Measurement results are evaluated annually by special transboundary committees and trend analyses are made available every fifth year. Accidental pollution spills which might have transboundary impacts are subject to special information-sharing requirements. Data-reporting tables are included.

1. Introduction

The River Tisza is the longest (977 km) tributary river of the River Danube, having the largest drainage basin as well. It drains water from the eastern parts of the Carpathian Basin and discharges into the Danube. The drainage basin area is 157,200 km^2 amounting to 20% of the total catchment area of Danube.

The present paper deals with the Upper Tisza river upstream of Tokaj (river kilometer, rkm 549) as shown in Figure 1. The catchment area of this upper basin is 49,500 km^2, which is shared by Ukraine, Romania, Slovakia and Hungary. Main tributaries of the River Tisza in this section are the Túr, the Szamos, the Kraszna and the Bodrog rivers.

The flow regime of the Upper Tisza and its tributaries is extremely varying. Annual total flow fluctuates with a ratio of 1:4, while the ratio of lowest to the highest flow can be as high as 1:70 within a year.

2. Present data sharing practices in the Upper Tisza River Basin

2.1. EXCHANGE OF HYDROLOGICAL AND HYDROMETEOROLOGICAL DATA

One of the characteristic features of the extremely varying flow regime of the Upper Tisza is the discharge of substantial flood flows.

I. Lyons Murphy (ed.), Protecting Danube River Basin Resources, 183–187.

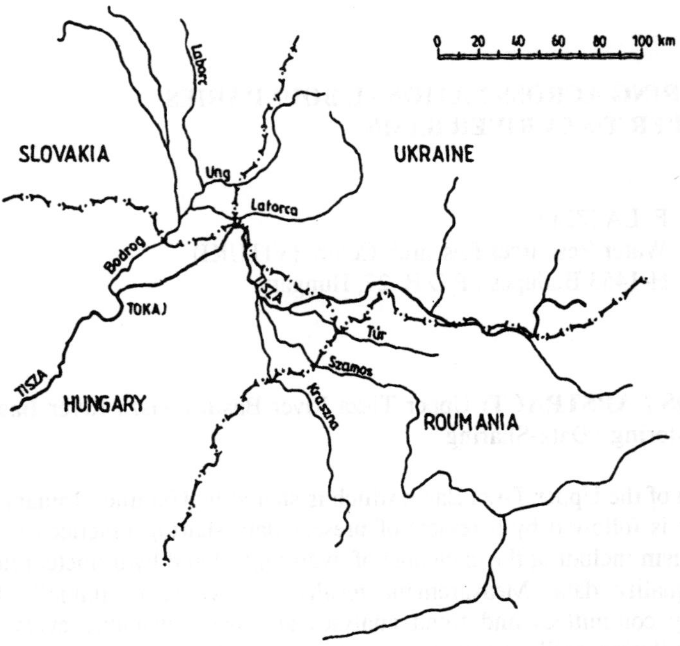

Figure 1. The Upper Tisza River Basin

From the point of view of flood defense the timely availability of basic hydrological and hydrometeorological data is extremely important for Hungary. The times of travel of flood peaks to the Hungarian border are as follows: Upper Tisza 20-30 h, Túr 20-30 h, Szamos 35-45 h, Kraszna 20-30 h, Bodrog 48-72 h.

At normal flow conditions the exchange of hydrological and hydrometeorological data is carried out through the national hydrographic centres (Ukraine: Ushgorod; Romania: Bucharest; Slovakia: Bratislava; Hungary: Budapest) on a daily basis. The number of hydrological and hydrometeorological stations involved in this data transfer are: 23 in Ukraine, 15 in Romania, 14 in Slovakia and 23 in Hungary.

In flood periods the rate of exchanging data accelerates, which means that information is exchanged several times in a day and there is a direct data transfer among the regional centres.

A special feature of data transfer in the flood periods is that Hungary transfers precipitation data measured by radar to Ukraine and Romania.

2.2 EXCHANGE OF WATER QUALITY DATA

Bilateral transboundary water committees in the Upper Tisza Basin had started to organize regular joint water quality investigations, water quality evaluation and the joint measures against accidental pollution events in the 1960s and seventies.

At present regular water quality investigations are carried out jointly with neighboring countries at the transboundary stations shown in Table 1.

TABLE 1. Sampling sites of bilateral water quality investigations of the Hungarian transboundary river sections of the Upper Tisza and its tributaries

River	Sampling station	Countries involved
Tisza	Tiszabecs	Ukraine-Hungary
Tisza	Záhony	Ukraine-Hungary
Túr	Kishódos	Romania-Hungary
Szamos	Csenger	Romania-Hungary
Kraszna	Mérk	Romania-Hungary
Bodrog	Fels_berecki	Slovakia-Hungary
Ronyva	Sátoraljaújhely	Slovakia-Hungary

Water quality parameters investigated are shown in Table 2.

TABLE 2. The parameters monitored according to bilateral agreements in the Upper Tisza River and in its tributaries at the Hungarian border sections

Parameter	River		
	Bodrog, Ronyva	Tisza	Túr, Szamos, Kraszna
Dissolved oxygen (DO)	+	+	+
Dissolved oxygen saturation	+	+	
BOD_5	+	+	+
COD_p	+	+	+
COD_d	+	+	
Ammonium	+	+	+
Nitrite	+	+	+
Nitrate	+	+	+
Total nitrogen	+	+	
Orto-phosphate	+	+	+
Total phosphorus	+	+	
Zinc	+	+	
Mercury	+	+	
Cadmium	+	+	

TABLE 2 - Continued

Parameter	River		
	Bodrog, Ronyva	Tisza	Túr, Szamos, Kraszna
Chromium	+	+	
Nickel		+	
Lead	+	+	
Copper	+	+	
Iron	+	+	+
Manganese	+	+	+
Water temperature	+	+	+
pH	+	+	+
Electric conductivity	+	+	
Total dissolved solids	+	+	+
Suspended solids	+	+	
Total hardness	+	+	
Chloride	+		
Sulphate	+		
Phenol index	+	+	+
Anionactive surfactants	+	+	+
Oil	+	+	+
Coliform bacteria	+	+	
Saprobity index	+	+	
Chlorophyll-a		+	+
Gross beta activity		+	
Cesium-137		+	
Strontium-90		+	
Tritium		+	
Water discharge			+

At the Tisza river stations Ukraine and Hungary carry out annually 8 investigations each, while the number of joint sampling is 4. In the rivers Bodrog and Ronyva two co-operating Slovak and Hungarian laboratories carry out 12 joint water quality investigations annually.

In the rivers Túr, Szamos and Kraszna annually 12 harmonized transboundary river investigations are carried out by each of the co-operating Romanian and Hungarian partners.

Sampling, sample treatment and analyses are being carried out by the co-operating laboratories on the basis of harmonized methodologies. A further element of analytical quality control is the intercalibration programme of the laboratories involved.

Measurement results are evaluated annually by the transboundary committees. Trend analyses are made in every fifth year.

In the case of accidental pollution incidents which are likely to have transboundary effects are subject to regulated transfer of information. Upstream countries transmit information to organizations of the downstream country at the time of observation of the accident, site and type of the pollution incident and on the kind and quantity of the pollutant involved. If additional information is available on river flow, velocity and concentration of the pollutant(s) involved they are also sent to the downstream country.

In addition to the bilateral exchange of data a preparatory study for investment, financed by EBRD, has also contributed to the exchange of environmental data in the Upper Tisza Region. In the framework of this programme the consulting agency has collected, with the assistance of local organizations, the available environmental data of the entire Bodrog river catchment and of the Upper Tisza Basin upstream of the confluence with Bodrog.

DISCUSSION FOLLOW-UP

With regard to whether there was harmonisation of methodology and analytical methods between Ukraine, Romania, and Hungary, the reply (Negelescu) was that they were similar, although usually it is just terminology that divides them. There is a general agreement on methodology and technical practices (Péter Literáthy, Hungary).

With respect to bio-physical methods it is very limited at this time, although it will become very important in the future. We need to discuss and agree on methods in all respects so as to avoid conflict. About heavy metals we have to weigh their significance, as well as the significance of dissolved forms and total concentrations (Varduca).

CROSSBORDER MONITORING OF WATER QUALITY IN SLOVENIA

D.SC. PROF. MITJA BRILLY
University of Ljubljana
Faculty of Civil Engineering and Survey
Ljubljana, Slovenia

D.SC. JASNA GRBOVIČ
Ministry of Environment and Physical Planning
Hydrometeorological Institute
Ljubljana, Slovenia

D.SC. META GORIŠEK
Ministry of Environment and Physical Planning
Ljubljana, Slovenia

KEYWORDS / ABSTRACT: International Water Quality Data Monitoring / Drava River Basin / Slovenia

Monitoring water quality data in Slovenia are part of the proposed environmental protection information system of the National Hydrometeorological Institute and will be both national and international in context. The database format is described as well as the legislative and document background of agreements to share monitoring data. International monitoring and data-sharing procedures are described in some detail. The importance of long-term monitoring data for the resolution of transboundary water quality issues is stressed.

1. Introduction

The state information system on environmental protection in Slovenia is currently a function of the Department of Planning in the Ministry of the Environment and Physical Planning. The task of organising the database for all environment monitoring has been assigned to the Hydrometeorological Institute of Slovenia. The data on monitoring water quality in Slovenia form part of the proposed environmental protection information system of the National Hydrometeorological Institute and are envisaged as both a data base of the national environment information system and the international environmental system, based on transboudary agreements and the international conventions.

I. Lyons Murphy (ed.), Protecting Danube River Basin Resources, 189–198.

In creating the integrated data base for water quality monitoring the existing software and hardware in the institutions which perform the water quality monitoring have been taken into account. A unified record form for parameters in the integrated data base has been defined which is now followed by all institutions participating in the monitoring. The following is the database format: consecutive numbers of fields, representing column labels, followed by strings of data for all measurements executed in a defined period. Columns follow one another in prescribed order, as defined in individual segments for each type of water individually. In the tables, which follow, each individual field, its type, field length, and the number of decimal places are set forth.

The sequence of parameters and length of individual records in any of files must not change. Should the need to increase the number of parameters or to upgrade the data base arise, new parameters will be added at the end of the file; any omitted parameters will remain in the record.

1.2. VALID REGULATIONS CONCERNING WATER QUALITY IN SLOVENIA

Valid regulations concerning water quality in Slovenia consist of legal acts of the former Yugoslavia recognised by the Republic of Slovenia after independence, legislation of Slovenia adopted before and after independence and the international conventions ratified by Slovenia.

1.2.1. *Legislation*

Legislation which applies to water quality monitoring and data sharing includes:

- Provision of classification of inter republican waters, transboundary waters and waters of the coastal sea of Yugoslavia 1978 (official journal of the SFR Yugosalvia, no 6/78);
- Decree on maximum admissible concentrations of radio nuclides and hazardous substances inter republican water courses, transboundary waters and waters of the coastal sea of Yugoslavia (official journal of the SFR Yugosalvia, no 8/78);
- Water act 1981 (official journal of the R Slovenia, no 38/81);
- Expert ruling on methodology for investigation of quantitative and qualitative changes in wastewater (official journal of the R Slovenia, no 4/85);
- The regulation on methods of sampling and laboratory analyses of drinking water 1987
- (official journal of the SFR Yugosalvia, no 33/87);
- The environmental protection act of R Slovenia (Ministry of Environment and Physical Planning, June 1993).

1.1.2. *International agreements*

The former Yugoslavia federation had signed agreements with the following neighbouring countries:

Republic of Austria

- Agreement on settling water management issues related to the transboundary course of the Mura river
- Agreement on settling water management issues of the Drava river

Republic of Italy

- Agreement on joint actions in protecting the water of the Adriatic Sea
- Agreement on the work of the commission for water management (continental waters)

Republic of Hungary

- Agreement on settling water management issues of the Drava river and the Danube river

The Republic of Slovenia has in 1992 on the basis of succession adopted additional measures:

- Agreement on water management issues of the Mura river (Austria)
- Agreement on water management issues of the Drava river (Austria)
- Agreement on the performance of the Slovene - Italian joint commission for water management (continental waters) (Italy)

1.1.3. *Conventions*

The Republic of Slovenia has on the basis of succession signed some of the most important conventions:

- Convention for the protection of the Mediterranean Sea against pollution and Protocol for the prevention of pollution of the Mediterranean Sea by dumping from ships and aircraft with oil and other noxious substances in cases of emergency and pollution from land based sources
- Protocol concerning Mediterranean specially protected areas
- Convention relating to intervention on the high seas in cases of oil pollution by substances other than oil
- Convention on the Protection and Use of Transboundary Watercourses and International Lakes (signed 1992 in Helsinki)
- Convention on Co-operation for the Protection and Sustainable Use of the Danube River (signed 1994 in Sofia) and Declaration of the Environmental Protection of the Danube River, that was based on this Convention and supports the Strategic Action Plan (SAP) of the Danube Countries

2. Overview of water quality in Slovenia

The Hydrometeorological Institute of Slovenia (HMIS) was established in 1947 and is responsible for water and air monitoring in R Slovenia since then. In the last five years, since Slovenia became an independent country, HMIS has been a public governmental organisation within the Ministry of Environment and Physical Planning. The basis of its activity is The Environment Protection Act of June 1993.

The main groups of activities carried out by HMIS in eight departments, are as follows:

- environmental monitoring including formation, maintenance and operation of monitoring;
- analysis and forecasting of meteorological, hydrological and ecological conditions and phenomena;
- study of weather processes, water regime and climate and their impact on different economical and other activities;
- study of physical and chemical properties of air and water related to environmental conditions and sustainable development.

The Hydrometeorological Institute of Slovenia is responsible for immission monitoring, laboratory tests and data collection (obligated by the Environment Protection Act of June 1993).

2.1. SURFACE WATER QUALITY MONITORING IN SLOVENIA

Monitoring water quality encompasses monitoring of springs and ground water, monitoring of flowing waters, monitoring of lakes and monitoring of the sea.

Monitoring of flowing waters and lakes is conducted by the Hydrometeorological Institute of Slovenia itself while the others are contracted to the following institutions:

- Institute of Biology at the University of Ljubljana
- Marine Biological Station-Piran, Institute of Biology at the University of Ljubljana
- Environmental Protection Institute - Public Health Institute, Maribor
- Public Health Institute, Novo Mesto
- Institute of Public Health of the Republic of Slovenia

All information is brought together in the Hydrometeorological Institute of Slovenia computer centre in an integrated data base, which was organised in 1992. Monitoring of flowing waters in Slovenia has been performed since 1965 when the sampling points (locations) network was composed of 20 locations while today are 110. Monitoring of flowing waters was established according to the methodology which is recommended by the World Health Organisation (WHO) and the World Meteorological Organisation (WMO). Water quality is defined on the basis of the results of chemical, physical, bacteriological and saprobiological analyses. Final evaluation of water quality is determined in close co-operation of all experts who took part in the monitoring program and schematically shown on the colour map for the current year. Monitoring of water quality is performed by the Hydrometeorological Institute of Slovenia, Department for the Environment. A comprehensive report on the results contains all data for the current year and a brief outline since 1990 (BILTEN) of the situation and of changes in quality as compared to previous years.

The most important criterion applied in determining the sampling frequency and the necessary extent of analyses was the possible influence of surface water on groundwater and major springs, both used for water supply. Sampling is done in various seasons of the year, preferably at low to mean low discharges. Samples for all types of analyses at one location are taken simultaneously. Biological samples are taken twice a year, once in summer and once in winter, while physical and chemical analyses are performed several times a year for each sampling site.

The water samples regularly include the measured values of fifty different basic physicochemical indicators; occasionally contents mineral oils in water, contents of 7 metals in water, suspended solids and in fluvial sediment, contents of organic compounds (phenols, pesticides, polycyclic aromatic hydrocarbons, polychlorinated biphenyl's, extracted organohalogen compounds, adsorbed organochalogen compounds) in water and in fluvial sediment. The biological samples include a list of all those present at the sampling site and their density. *Our interpretation* of causal relationships based on *real biological properties* relating to the environment rather than just on statistical interpretations. Our classifications based on an examination of periphyton and makrozoobenthos present in water at location. A basis for our biological evaluation of the water quality of running waters are the as complete as possible species determination of organisms composing the communities, their semiquantitative determination (abundance scale 1-3-5) as well as the knowledge of their autecology. this is saprobic system, which classifies running waters into 4 classes:

1st -	I.	oligo-saprobic, in which the water is *unpolluted,*
2nd -	II.	beta-mesosaprobic, in which the water is *slightly polluted,*
$3r^{d}$ -	III.	alfa-mesosaprobic, in which the water is *polluted,*
4th -	IV.	poly-saprobic, in which the water is *hardly polluted.*

In routine biological monitoring, we try to detect significant changes or trends in watercourses, not minor fluctuations that are quickly dampened.

2.2. MONITORING OF LONG-TERM WATER QUALITY AT DRAVOGRAD

We have been conducting, jointly with Austrian experts, hydrobiological and hydrochemical analyses at Dravograd for about thirty years. Hydrobiological samples are collected in lower part of the accumulation, on the border between former Yugoslavia, today Slovenia and Austria, whereas the samples for physical and chemical analyses are collected in the profile Tribej in Austria. Hydrobiological samples collected with a plankton net and bottom corer are examined alive in a field laboratory, then fixed and thoroughly examined in the laboratories in Ljubljana and Klagenfurt. Physical measurements and chemical analyses of water carried out every 4 hours in a 24-hour period; some of parameters are determined in a field laboratory and the others in laboratories in Ljubljana and Klagenfurt.

During the thirty year observation period the river Drava in the profile Tribej maintained second quality class.

3. Water quality monitoring of transboundary impact

Co-operation between Slovenia and Austria has a long tradition (see Figure 1). Commissions between the Republic Austria and the formal Yugoslavia to implement monitoring were established in 1954 for the Drava river and in 1955 for the Mura river. Since 1965 water quality monitoring and analyses have been performed on the Mura in Spielfeld, on the border with Austria. During this time the extent of the analyses and the frequency of sampling were changing often. The water quality of the river Mura varied. According to the results of chemical and physical analyses the Mura in Spielfeld was in the 4th quality class in the 1965-1980 period, in the 3rd and 4th quality class in the 1980-1988 period and in the 3rd quality class in the 1988-1991 period. From 1992 until now the Mura in Spielfeld shows between the 2nd and the 3rd quality class (see Figure 2). Hydrobiological analyses have shown different quality classes compared to chemical and physical ones. This is due to different methodologies. Even though the trend of both types of the analyses is the same.

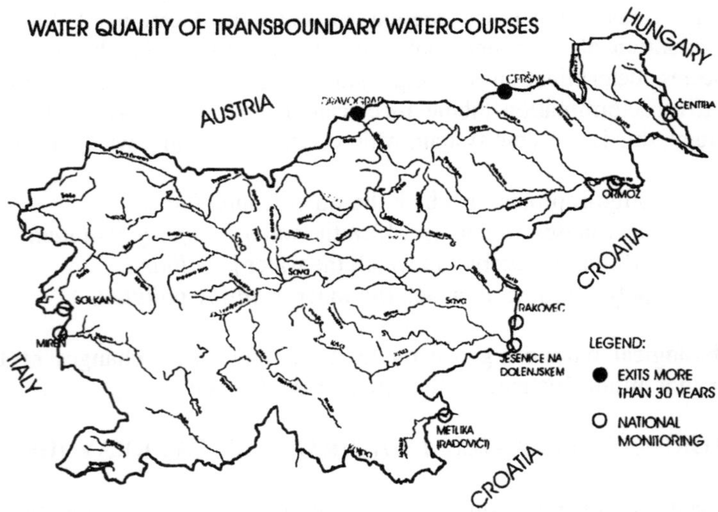

Figure 1. Map of Slovenia with *transboundary* monitoring profiles on the boundary rivers

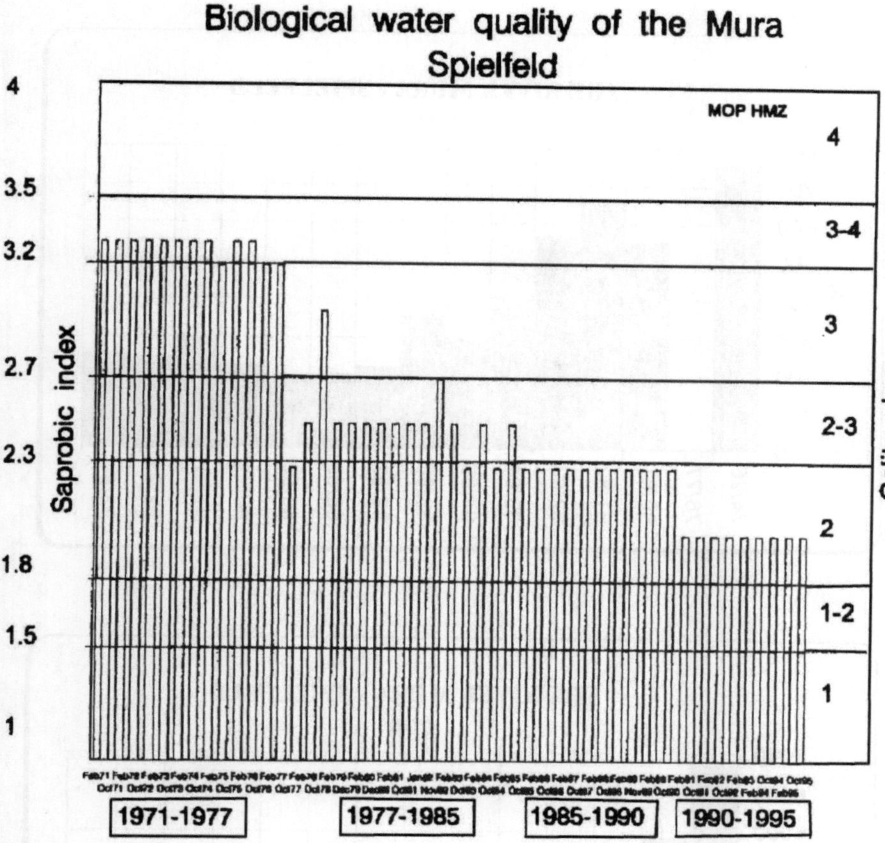

Figure 2. The variations of the saprobic index of the Mura river at Spielfeld in the period of 1971-1995

The improvement of the water quality of the river Mura on the border profile in Austria is connected to treatment actions taken in Austria. The significant improvement of the water quality of the Mura river cross section at Spielfeld after 1984 is confirmed by lower chemical oxygen demand (COD) and biological oxygen demand (see Figure 3) and is caused by the Austrian Government regulation measures of the outflow in the paper mill industry. The water quality of the Mura river improved from the 4th to the 3rd class.

The circumstances of the water quality improvement of the Drava river are quite different (see Figure 4). The chemical oxygen demand (COD) is significantly lower but from the biological point of view the water quality of the Drava river is almost the same. The lower industry emission from the Austrian industry to the Drava river did not influence the biological system of the river to any great extent. Rather the change has come about because of the construction of the chain of hydroelectric power plants.

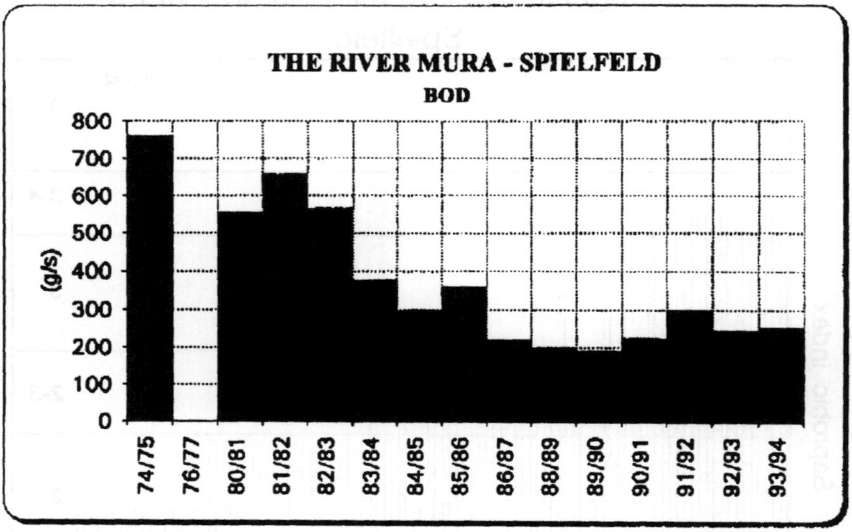

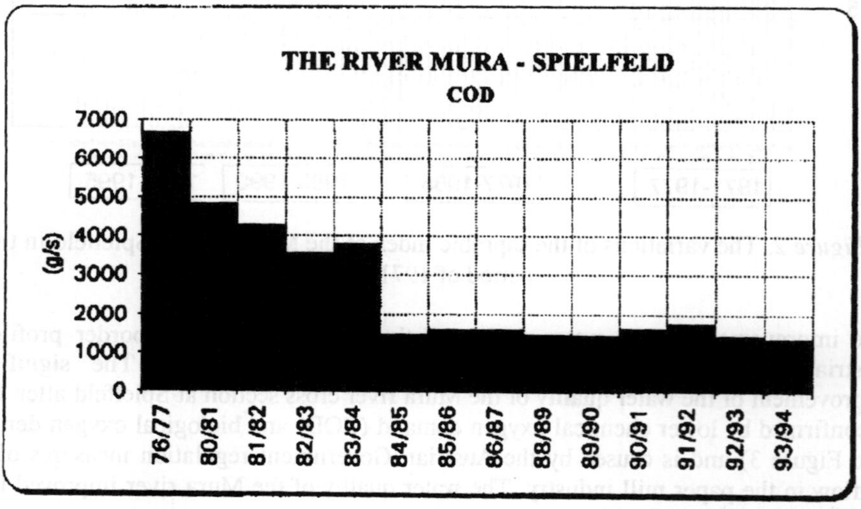

Figure 3. The measurements of BOD and COD at Spielfeld profile, the Mura river

4. Conclusions

The availability of long-term monitoring data are very important for transboundary water quality issues. The sampling data of only few years duration cannot be the input data for decision making and negotiations about the common goals and the performance of the common water management and environmental protection strategy of neighbouring countries.

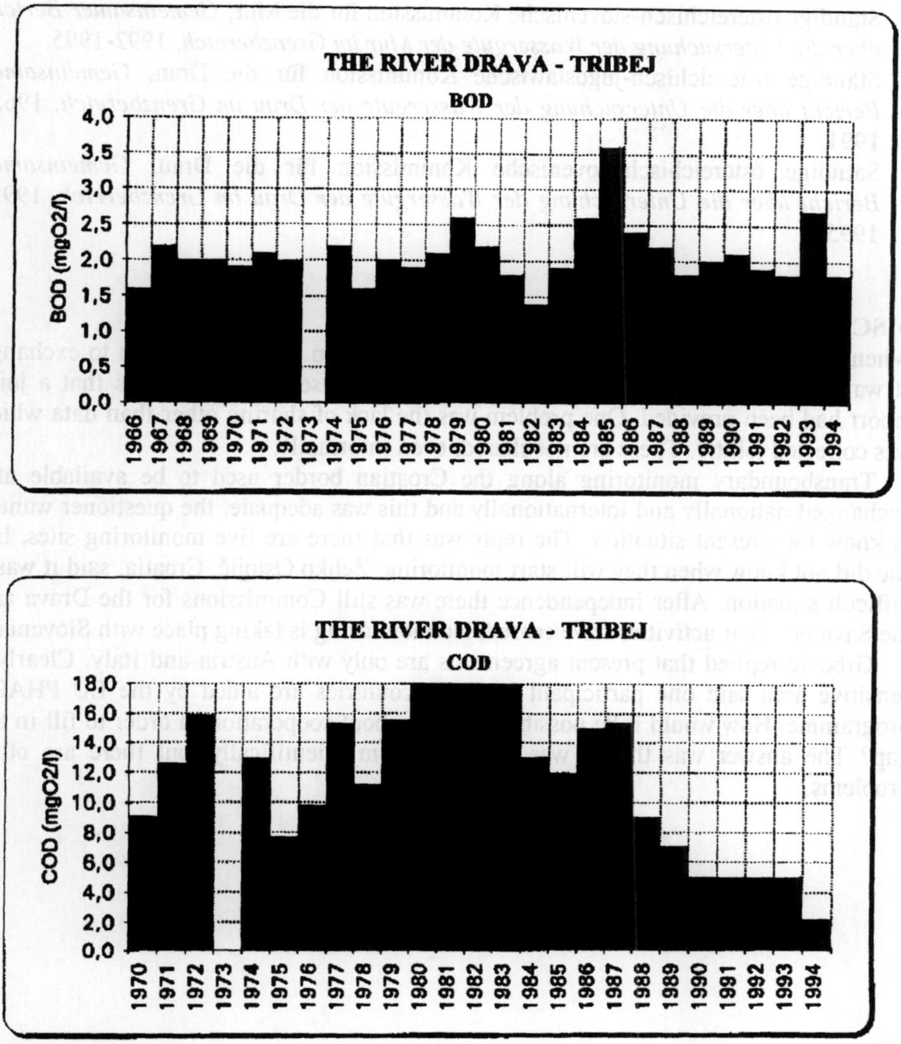

Figure 4. The measurements of BOD and COD at Tribej profile, the Drava river

References

1. Zupan, M. et al., Water Quality in Slovenia Year 1993. *Published by Ministry of the Environment and Physical Planning*, Hydrometeorological Institute of Slovenia
2. Ständige östereichisch-jugoslawische Kommission für die Mur, *Gemeinsamer Bericht über die Untersuchung der Wassergüte der Mur im Grenzbereich*, 1965-1991.

3. Ständige östereichisch-slovenische Kommission für die Mur, *Gemeinsamer Bericht über die Untersuchung der Wassergüte der Mur im Grenzbereich*, 1992-1995.
4. Ständige östereichisch-jugoslawische Kommission für die Drau, *Gemeinsamer Bericht über die Untersuchung der Wassergüte der Drau im Grenzbereich*, 1965-1991.
5. Sätndige östereichisch-slovenische Kommission für die Drau, *Gemeinsamer Bericht über die Untersuchung der Wassergüte der Drau im Grenzbereich*, 1991-1995.

DISCUSSION

When asked if the authors were in a position to sum up issues in relation to exchange of water data including recommendations, the response by Grbovič was that a joint report had been provided. One problem was the lack of sharing other than data which was collected jointly. These are not shared, even informally.

Transboundary monitoring along the Croatian border used to be available and exchanged nationally and internationally and this was adequate, the questioner wanted to know the present situation. The reply was that there are five monitoring sites, but she did not know when they will start monitoring. Željko Ostojič, Croatia, said it was a difficult situation. After independence there was still Commissions for the Drava and the Sava but their activities have ceased. No monitoring is taking place with Slovenia.

Grbovič replied that present agreements are only with Austria and Italy. Clearly a sensitive area said one participant but both countries are aided by the EU PHARE programme. How would it be possible to bring about cooperation in order to fill in the gap? The answer was that it was not a problem scientifically, but there are other problems.

WATER QUALITY DATA SHARING ON THE BULGARIAN-ROMANIAN REACH OF THE DANUBE RIVER BASIN:

The need for harmonisation of legislation

ATANAS PASKALEV, Ph.D.
Bulgarian National Association on Water Quality
3 Oborishte, Str, 1504 Sofia, Bulgaria

KEY WORDS / ABSTRACT: Danube River Basin / Bulgaria / Romania / environmental legislation / water quality monitoring / water quality data-sharing

This paper describes the need to harmonise the environmental legislation of the East European countries (EEC) with that of the West European countries and the European Community so that water quality and other data can be shared among all countries. The progress made by the Environmental Programme for the Danube River Basin (EPDRB), the Danube River Protection Convention (Danube Convention), as well as National Action Plans are main supports for sustainable development in the region. United action across national boundaries is required. The boundary between Bulgaria and Romania in the lower part of the Danube river suggests the importance of creating and supporting harmonised and mutually relevant laws and regulations for the protection of the water quality of the rivers discharging into the Danube River from each country.

1. Introduction

Much research and many practical investigations into optimal waste treatment have attempted to set forth management principles for optimal use of Danube River Basin resources. These efforts include as their goals a strategy for the solution of water pollution problems in the Romanian-Bulgarian area of the Danube and the restoration of a healthy aquatic life in the entire Danube-Black Sea system.

The Danube is the recipient of a large amount of nutrients and toxic pollutants, delivered from point and non-point pollution sources. In addition to domestic wastes, there is much waste from industrial, transport and agricultural activities. These substances pollute surface and groundwater in the area through waste water discharges, surface runoff, atmospheric fallout and leaching from solid waste disposal sites. In addition to these chronic pollution inputs the water quality along the river is influenced by accidental pollution, man-made structures and natural processes. Consequently the pollution from the Danube catchment to the Black Sea is responsible for eutrophication in the north-western sea areas [8].

I. Lyons Murphy (ed.), Protecting Danube River Basin Resources, 199–207.

All these facts mean that international collaboration among the countries in the catchment areas is essential for continuous working out of the details of enforcement and further refinement of existing conventions. Important aspects of this collaboration include the Bucharest Declaration, the Environmental Programme for the Danube River Basin, the Black Sea Convention and the Danube River Protection Convention. The enforcement of each agreement, as well as data and information exchange, are closely related to the harmonisation of member countries' legislation. The need for environmental legislation and standards harmonisation is especially emphasised in the Danube Convention in Article 9 [1] and in the Strategic Action Plan for the Danube River Basin [11].

2. Some general concepts

2.1. THE BUCHAREST DECLARATION

During the period of centralised planning, the Central and East European countries did not develop comprehensive environmental protection policies and thus contributed to the degradation of the river environment. Legal standards for environmental quality were often neither enforced nor enforceable [11]. For example the Bucharest Declaration on Water Management of the Danube, signed in 1985 states that riparian countries will protect the Danube against pollution with special regard to radioactive and dangerous substances in order to gradually decrease the level of pollution; and that they will establish national systems for the monitoring of waste waters discharged into the river system. The Declaration, however, is a statement of intent rather than a legally binding document.

Working groups on water quality, flood information and forecasts and water balance were established within the framework of the Declaration. In the ten years following its adoption, a monitoring programme has been instituted and 11 water quality stations and 13 flow stations placed on the river. Sampling and field measurements have been carried out on a bilateral basis by the border countries concerned. Expansion of the sampling and field measurements has been suggested. There are thousands of surface water monitoring stations and tens of thousands of groundwater monitoring stations operating under national jurisdiction throughout the basin. Despite this the monitoring system put in place under the protocol of the Bucharest Declaration has not been sufficient and experiences differences in monitoring technology. Greater standardisation of monitoring, laboratory analysis and information management within riparian states are still required to achieve the objectives of the Declaration.

The lessons to be learned from the Bucharest Declaration in respect to the monitoring include the need for:

- extension of the sampling programme to heavy metals and selected organic micropollutants in sediments and biota with less frequency than that applied to water;
- revision of the location of sampling sites;

- revision of sampling and analytical methods;
- incorporation of biomonitoring, at least for macrozoobenthos, and
- enforcement training programs and quality control measures and the continuation of the study of compatibility of interlaboratory (intercalibration) analysis as well as of improved, more responsive exchange of data and information between all countries.

2.2. MONITORING PROGRAMMES

There is a pattern of insufficient co-ordination between water quality measurement and hydrological measurement, resulting in a major deficiency in the calculation of loading, mass balances and the import and export of pollutants. An important priority is the co-ordination of the activities so that effective and efficient monitoring could be developed. Quantification of the river loading is viewed as one of the most important aspects of a monitoring system, and this importance is underlined by the fact that so many countries are involved. Problems of load quantification are often due in part to a lack of technical equipment and resources.

National quality monitoring programs in the Danube countries, including Bulgaria and Romania, have provided and continue to provide a vast amount of data on water quality. Results have not been uniformly comparable due to differences in sampling and analytical techniques, and the lack of quality control. Monitoring of toxic pollutants was random concerning the particular compounds, sample matrix, i.e., water, sediment or biota, sampling sites and position, and frequency.

Recognition of these problems was the occasion for Article 9 of the Danube Convention that "the Contracting Parties shall agree upon monitoring points, river quality characteristics and pollution parameters regularly to be evaluated for the Danube River with a sufficient frequency taking into account the ecological and hydrological character of the watercourse concerned as well as typical emissions of pollutants discharged within the respective catchment area, as well as to develop joint or harmonised methods for monitoring and assessment of waste water discharges including processing, evaluation and documentation of data taking into account the branch-specific approach of emission limitation."

2.3. PRESENT NEED FOR COOPERATIVE ACTION

It appears that both Bulgaria and Romania, like other Central and East European countries responsible for the process of economic, political and social transition, to reap full benefit from closer links with Organisation for Economic Co-operation and Development (OECD) countries, will have to harmonise their environmental standards and legislation with those of the West, and in particular with those of the European Community. Many political, economic and environmental factors are driving harmonisation forward.

Both Romania and Bulgaria have formally submitted the required documents for full membership in the European Union. Harmonisation with EU Council directives is therefore an important driving force of the harmonisation process and sets a clear target for its successful conclusion. Signing the European Agreements is a commitment to action by the Central and East European countries. Potential investors may be discouraged by the complexity and lack of transparency in existing environmental legislation in Central and Eastern Europe which not only differ from that of the OECD countries, but also varies from country to country. Harmonisation could contribute to enhancing the transparency of requirements and procedures [5].

Central and Eastern European exporters would also benefit from harmonisation, since growing awareness of environmental issues by consumers in the OECD countries is likely to cause the imposition of stringent environmental standards on imports from the region. All these factors should help to ensure the future development and sharing of water quality data between Bulgaria and Romania.

3. Water quality in Bulgaria and Romania

Bulgaria shares 471 km of the Danube river with Romania. The total population in the Bulgarian part of the Basin is 4.35 million inhabitants. The contribution of its tributaries to the Danube is limited, around 6.2 billion m^3 per year, less than 4% of the total Danube runoff. Romania has a population of approximately 22.76 million. Its tributaries contribute to Danube waters approximately 14% of Danube flow.

The purification capacity in this part of the Danube for oxygen depletion substances is high and is able to cope with existing BOD (COD) discharges into the main river. The tributaries are much more polluted than the Danube itself; in many instances the concentration of pollutants greatly exceeds all acceptable standards. A number of tributaries experience a very high concentration of BOD. This reduces the availability of water and results in a very high treatment cost of municipal and industrial water supplies, as well as causing significant ecological damage and presenting a risk to public health. The last study (undertaken by Haskoning [2]) on identifying the target sectors that contribute to the threatening of key functions through the discharge of priority pollutants into surface water shows that population mainly contributes to bacteria and viruses, nutrients, metals and oil pollution.

3.1. INDUSTRIAL POLLUTION

A great part of this pollution comes from industry. Industrial waste waters are 30% to 60% of the total waste water treated in the municipal waste water treatment plants (WWTP) in CEE countries. Generally, WWTPs have an extended oxygen capacity and indirect discharges with a higher content of biodegradable substances (expressed as BOD or COD as well as SS) and do not present a threat in their normal operation.

Monitoring experience in these countries (especially along the Danube downstream) shows that peak loads of mazut and other oils and industrial pollutants often occur and that this phenomenon may interfere with the normal operation of WWTPs. The best and often the only tool for preventing such anomalies is the enforcement of proper legislation. Standards are not obeyed, and compliance with existing regulations not required. Temporary (less stringent) regulations have been worked out during recent years. It is not yet clear which branches of industry will continue to exist and therefore it is difficult to predict future pollution loading.

The EU Member States have clear and sound rules, regulating the pre-treatment of the industrial waste waters entering the sewerage systems and municipal waste water treatment plants [1].

3.2. EU REQUIREMENTS

While Bulgaria and Romania have submitted documents for full membership in the EU it is a question as to whether governments and the appropriate bodies know what activities should be done and what resources should be allocated for the problems of the waste water discharged from WWTPs.

The urban waste water Directive of the Commission of the EU or Directive 91/271/EEC May 21, 1991 would probably qualify the Danube as sensitive, which means that the riparian countries should build their sewerage collecting systems and WWTPs and apply nutrient removal for the major waste water treatment plants by the year 2005. To some extent this is in conflict with the environmental strategy of the Central and Eastern European countries, especially in the part concerning nutrient removal. So it is expected that downstream countries will need a longer period extending beyond 2004 to comply with the Directive to qualify for EU membership.

4. Some problems of Waste Water Treatment Plants in Bulgaria and Romania

Many studies carried out and many papers have been published for the last years, discussing the Waste Water Treatment Plants (WWTPs) in the countries of Central and Eastern Europe (CEE), situated in the Danube river basin. The publications of Somlyódy [9], [10], Henze and Odegard [6], Grau [7] and many other European authors outline the frameworks for the elaboration of an Environment Action Plan for Central and Eastern Europe [11]. This information helped to finalise the Strategic Action Plan (SAP) for the Danube River Basin. Now, at the beginning of 1996 the Danube Basin countries work upon their National Action Plans to help implement the SAP. There will be many opportunities for each country to bring WWTPs and other forms of waste management in harmony with those in western Europe.

Most WWTPs in operation in Bulgaria and Romania need some form of reconstruction, up-grading and enlargement. At the same time there are a significant number of plants which are not fully utilised for lack of supporting sewer system, interceptor sewers, or pumping stations. Many treatment plants are compromised by poor design, low quality outdated equipment with high energy demand especially for aeration basin, inadequate control and monitoring system, inadequate maintenance, and poor operation. Most are significantly over sized in their design or unloaded due to past practices in forecasting treatment.

The lack of adequate industrial pre-treatment is a problem in both countries. Industrial waste water often contains toxic or other undesirable components, such as heavy metals, oils, toxic organic compounds, extremely acid or alkaline wastes and so forth. The gap widens further when taking into account the treatment of such sludge, when there is no place for its disposal.

The various factors degrading the quality of sludge prevent its use as fertiliser or in energy production. Often sludge find its way back into the environment, reversing the environmental benefit intended from waste water treatment in the first place. In this respect there is a paradox in Bulgaria: in order to protect environment, the Council of Ministers issued a decree N153 dated August 6, 1993 concerning "Collecting, Transporting, Preserving and Making harmless the Toxic Wastes". The same decree states that the WWTPs' sludge are "hazardous wastes" until the contrary is not proved. Hence the reuse of sludge as fertiliser will be a long and difficult process.

The urban WWTPs in both countries require huge investments and time so as to achieve the appropriate operating level. According to the World Bank report [10] if we assume that 1.5% of the GNP could be spent on sewers and wastewater treatment plants, the calculated average period of meet EU water quality standards for Bulgaria is 113 years and for Romania 58 years. At first sight these data seem exaggerated, but the real situation in the Danube river downstream countries bears testimony that many years will pass, before things get better. In general, the existing WWTPs and conveying collectors were built in the 1970s and 1980s. These installations are in the second stage of their operation now. They will need renovation, reconstruction and upgrading in the near future. All these matters require huge investments and a long period of time.

5. Some advantages and obstacles in harmonisation of environmental legislation

Institutional reform is a prerequisite for implementing harmonisation strategies. It is clear that the cost is the major constraint for the speed at which Bulgaria can move towards implementing EU environmental standards. Parliaments throughout Eastern Europe have not much time to spend on environmental legislation since they have to pass many other laws concerning the entire state's activity. As for the water legislation, it is now under consideration in the Bulgarian parliament and will be passed soon.

In both countries the principle "polluter pays" is priority, but whilst there are common principles, the level of detail and degree of environment of the law varies between both countries. In both countries licensing systems are largely in place but they have not been applied consistently. As a result a large proportion of untreated industrial water is often discharged directly into surface waters. Law enforcement is a result of the ongoing institutional changes mentioned earlier. Within this enforcement and legislative void, industrial restructuring and privatisation continue but with great need for well-designed policies and strong enforcement.

5.1. DATA STANDARDIZATION

It is important to note that the problem of lack of standardised data is more serious than just the classification systems, for the actual methods of sampling, testing, analysis, and documentation also vary widely and even seemingly compatible results way have been obtained in very different ways. It is strongly felt that the lack of a standardised basin-wide monitoring system is one of the most important factors constraining environmental enhancement in the region. The comparison of data between the two countries can very well contain errors. The need of harmonisation is obvious.

The analysis of STAS 4 706/1988 made by Dr. Varduca [12] is a progress for the Romanian side. According to this analysis the harmonisation of results is necessitated at world level. This was especially imposed by transboundary control which led on the one hand to the promotion of generally applied standard analytical methods (hardware for data acquisition) and on the other hand to setting-up of the resulting inter-calibration requirement from both acquisition and processing procedures point of view. Therefore, it is expected that the Romanian regulation will be changed soon. The changes will be directed towards combining the opportunities for water use both as a resource and an environmental component. Similar regulation is being elaborated in Bulgaria and it is expected to be enforced soon.

Cost recovery and tariff in the water supply and sewerage sector have an important impact on the quality and quantity of effluent which is discharged into the environment. Traditionally, water supply and sewerage were provided by the government as a social service at a nominal charge which remained unchanged for years. This led to excessive and wasteful water consumption practices by both domestic and industrial consumers. The combination of inadequate cost recovery and increasing government budgetary constraints have led to:

- excessive water consumption and high sewage flows;
- historic over-investment in water supply and sewerage facilities;
- lack of incentives for industries to pre-treat effluent before discharge to municipal sewers;
- shortage of spare parts and other operating materials; and
- shortage of funds for regular preventive maintenance.

5.2. ENFORCEMENT OF LEGISLATION

In implementing the legislation, due regard must be paid to the practicalities of the economic situation. In particular, a degree of flexibility should be applied in determining the timing of the imposition of the new pollution fines to existing industries or other polluters. Prior to imposing the new charges in a particular case, it would be desirable, first, to carry out an Environment Impact Assessment - EIA (or audit) of the operation concerned; on the basis of the EIA, a realistic timetable could then be defined for the implementation of a pollution control investment programme at specific sites, through negotiation between the polluter and the Ministry of Environment.

This sort of approach seems preferable to the alternative of immediate application of the new, much higher penalties, which might result in severe financial problems for the polluter (or even bankruptcy), and would reduce his ability to finance the needed investment in pollution control. Monitoring and enforcement are key issues in pollution control and are the final step in ensuring that environmental standards are satisfied. Existing monitoring system is irregular with pollution charges and fines which are largely ineffective because they are either too low or not applied in all cases. This situation should be addressed as a matter of urgency if the government is to implement a credible environmental policy which the general public has confidence in and ensures a sustained positive reaction from the polluters themselves.

The key components to a realistic monitoring and enforcement programme are as follows:

- Plant specific discharge/pollution agreements which are renegotiated every 5 years, with defined targets for pollution reduction, plus agreed monitoring and reporting requirements. The statement of pollution charges/fines should include an allowance for inflation.
- Agreements should also require self-monitoring and reporting of effluent by the enterprise and make the reporting of false information a criminal offence.
- Agreed emergency action plan in the event of a major accident.
- An effective pollution monitoring programme linked to an environmental data management system.
- Regulation for monitoring frequency, the type of analysis to be used, the sampling method, the point of sampling and form and frequency of results reporting.
- Strict enforcement and collection of stipulated pollution charges and fines.
- No exemptions should be permitted and the MOE should be prepared to take firm legal action against offenders who persistently exceed the pollution targets.

It should be emphasised that monitoring needs to be carried out in a cost-effective manner which satisfies the objectives of the stated environmental policy and agreed pollution standards. In order to make a substantial impact, initial monitoring and enforcement activities should be focused on "hot spot" pollution (point and non-point). This would yield the most significant results, be cost-effective and demonstrate to the public and polluter alike that the Government is taking firm action. But true costs of the service are not known as the financial systems in both countries have not been established to identify the individual cost centres.

6. Conclusion

One of the first reports on the problems of harmonisation of the Danube River Basin countries' legislation was the Concept Paper prepared by WRc [13]. This paper emphasised that the harmonisation should follow previously chosen priorities. The WWTPs problem was involved in the short term strategic directions of the Strategic Action Plan for the Danube River Basin. According to Sub-Chapter 2.5.1. of the SAP, which has been adopted by both Bulgaria and Romania, in the short term monitoring and assessment methods should be stressed and the first phase of the joint monitoring water quality, quantity, sediments and riverine ecosystems put into operation.

Harmonised methods for transboundary impact assessment of pollution and damage to riverine region and the adoption of harmonised monitoring and assessment methods for waste water discharges should follow. Determination of methods and development of funding for harmonisation of all problems connected with WWTPs is justifiable and timely.

Some important considerations include the following:

- Harmonisation of the legislation in the field of urban waste waters depends on the co-ordination of activities related to: institutional arrangement and state responsibility; water acts and related acts; environmental standards and regulations; control of polluting industries and choice of technology; control of sewage effluents and discharges to sewers; charges and fines; monitoring and licensing; compliance and enforcement of cost-effective technology and river basin least cost approach.
- Harmonisation in waste water sector needs to be flexible, to be undertaken within a reasonable time frame, to be cost-effective and focused on priority problems.
- Because of lack of funds for investment in Bulgaria and in Romania and shortage of technical resources, the process of harmonisation will be a long one, taking probably decades.

By necessity, harmonisation should be a gradual process of convergence involving data collection and information and programme sharing for the effective management of Danube River Basin Resources.

References

1. Convention on Co-operation for the Protection and Sustainable Use of the Danube River, (1994) Austria.
2. Danube Integrated Environment Study, *Final report Haskoning* (1994), Netherlands
3. Danube River Basin Environmental Programme, Analysis and Synthesis of National Reviews, *Final Report*, (1994).
4. Environmental Action Plan for Central and Eastern Europe, (1993).
5. Environments in transition (1993), *The Environmental Bulletin of the EBRD*, Lucerne, Switzerland
6. Henze M., Odegaard H, An Analysis of Wastewater Treatment Strategies for Eastern and Central Europe (1994), *Proceeding of the IAWQ 17th Biannual International Conference*, 27 - 42.
7. Grau P. (1994), What next?, *Water Quality International* **4**, 29 - 32.
8. Mee L.D.(1992), The Black Sea in crisis a need for concerted international action, Ameba, 21, 278-286.
9. Somlyódy L., Quo Vadis Water Quality Management in Central and Eastern Europe? (1994), *Proceeding of the IAWQ 17th Biannual International Conference*, 1-15.
10. Somlyódy L. (1993), Looking Over the Environmental legacy, *Water Quality International* **4**, 17- 20.
11. Strategic Action Plan for Danube River Basin 1995 - 2005, *Environmental Programme for the Danube River Basin by the Task Force for the Programme*.
12. Varduca A., Marculescu V. (1993), Points of View Regarding Main Substation Element Necessary for Revising STAS 4706/1988, *Medivl Incowurator* **4**, N 2.
13. Zabel T. F., Buckland J. (1993), The Implications of Harmonisation in the Water Sector, *Concept Paper for the Workshop in Teplice*.

CONCLUSIONS AND RECOMMENDATIONS

1. Conclusions

1.1. INFORMATION AND DATA SYSTEM DEMONSTRATION

The demonstration provided an opportunity to review a range of information systems and databases, some of which were now operative with respect to the Danube; others were more general.

David Rodda, who served as team leader of the Environmental Programme for the Danube River Basin for the last four years stated in his paper that "[b]oth data and information are vital to effective regulation and future planning for the implementation of environmental protection measures. No management strategy is complete without an information strategy... an information strategy is emerging, but it is at an early stage." As information technology advances, data and information tend to increase in volume and complexity. These concepts guided the discussion of the final conclusions and recommendations.

1.2. A COMMITTEE AT WORK: MAJOR ISSUES

The following is a summary of the discussions which preceded the adoption of the conclusions and recommendations of the workshop on Thursday morning, 30 May 1996. The organising committee of the workshop had met on Sunday afternoon 26 May 1996 to discuss various aspects of the workshop. In particular they reviewed the proposed plan to appoint a committee to help draft the final recommendations of the workshop. It was decided that Dr. Tim Lack, of the EPDRB MLIM, should facilitate the adoption of the action strategy with the help of the drafting committee.

During the Monday afternoon workshop session Dr. Péter Bakonyi, co-director of the workshop, reviewed its goals and emphasised that their implementation would require the concerted effort of a small group. He also emphasised that they would be asked to come up with suggestions by Wednesday so that all participants would have an opportunity to discuss them and add input. Those who participated in the drafting committee session were: Tim Lack, Chairman, Ed Bellinger, Hans Liebscher, Atanas Paskalev, Jan Kubat, David Wangsness and Karl Schwaiger

A number of issues arose at the meeting of the drafting committing as well as the general session of the workshop participants as they considered the action strategy. These include (not necessarily in order of importance):

1. the group had concluded that the workshop was about data and information sharing rather than data-sharing alone; this was largely a question of terminology and there were many more significant issues;

I. Lyons Murphy (ed.), Protecting Danube River Basin Resources, 209–214.

2. the concept of the final use of the data should be paramount; such use is guided by international agreement but must be more than obligatory, it should have significance beyond that;
3. publication of data is very important; it has already been decided that the Danube International Commission must make monitoring data available, that is what the MLIM has been doing by establishing the TNMN; the countries themselves have agreed and will move forward with support for this and the DEFF; international determinands have been agreed to;
4. all of the data (explained by Tim Lack) should be compiled within three months; its aggregation should result in a report similar to those published now about the Rhine and the Elbe; we can have a Danube metadatabase which would show what is available and where it is available;
5. certain aspects of data collection still need research, e.g. the harmonisation of biological sampling;
6. there is a basic need to bring both transboundary information and data in a network together at one point to provide a single point of inquiry to meet all needs; a World Wide Web (www) application, a host organisation is needed for management and updating the system; Danube countries could provide home pages for their respective countries;
7. the system should make use of sources being maintained by others at their expense, therefore minimising costs.
8. user groups include responsible authorities, decision makers, research institutes, public, academicians, the non-governmental private sector, industrial and commercial organisations;
9. linkages should be developed within the system and with EU information systems

The detailed paragraph by paragraph review of the proposed recommendations settled a number of points regarding word usage, emphasis, order, etc. A lengthy discussion about the confidentiality of data and information was followed by agreement that only data and information that are officially approved for publication or public release should be placed on any system but that discretion belonged to the owners of the information whether they were the database owners, individual country governments or an international agency such as the Danube International Commission.

2. The Recommendations

RECOMMENDATIONS FOR FURTHER ACTIONS FOR IMPROVING THE EXCHANGE AND DISSEMINATION OF DATA AND INFORMATION ON THE DANUBE RIVER BASIN

Adopted 30 May 1996
Central European University
Budapest, Hungary

At the Nato Advanced Research Workshop "An International Data-Sharing Programme for the Effective Management of Danube River Basin Resources"

1 BACKGROUND

1.1 This draft proposal has been produced as an output of the North Atlantic Treaty Organisation (NATO) Advanced Research Workshop (ARW) "An International Data-Sharing Programme for the Effective Management of Danube River Basin Resources" (ARW: ENVIR.ARW 951007) in accordance with the requirements of the funding provided by the NATO Science Committee.

1.2 The workshop involved a selected group of experts from governments and institutes having responsibility for managing and disseminating information in the Danube River Basin supported by expertise from international river commissions and other institutes of Europe and North America.

1.3 The further actions recommended by the ARW represent the collective views expressed at the Workshop making reference to significant aspects of work so far undertaken by the Danube countries and set out in the provisions of the Danube River Protection Convention and the activities of the Environmental Programme for the Danube River Basin (EPDRB).

1.4 Thus, the Workshop took note of Article 12 of the Danube River Protection Convention, and Action Point 38 of the Strategic Action Plan of the Environmental Programme of the Danube River Basin. The latter states: "The sharing and exchange of data and information are fundamental to regional cooperation and to the understanding and solution of regional problems. The Danube countries should share information on, for example, sediment transport and erosion, nutrient balances, low flow and water quality control, legislation and integrated planning."

1.5 Data and information on the environmental state of the Danube River Basin have arisen, over time, in a number of formats with a variety of uses. Data and information cover such subjects as: legislation and standards, institutional organisation, names and contact points of key actors, NGO activity programmes and outputs, literature sources, research programme reports as well as water quality characteristics measured at a number of stations in the Danube and its tributaries.

1.6 Data and information exist in paper form (for example as reports and year books) and electronic databases (for example INFODANUBE, DANIS, Environmental Standards Database, Bucharest Declaration).

1.7 For water quality and related data, in accordance with agreements by Monitoring, Laboratory and Information Management Study Group of the EPDRB, data from the Trans National Monitoring Network (TNMN) stations will flow from the National Information Centres (NICs) to the International Commission for the Protection of the Danube River (Danube International Commission) at agreed frequencies and determinands. The current agreement is for river basin countries to exchange data via the Data Exchange File Format (DEFF) on diskette annually and to have a central point for the aggregation of the data in order to produce an annual report on the status of the Danube river basin. It is assumed that this central point will be the Danube International Commission although interim arrangements have been made.

1.8 In order to improve the broader dissemination of officially approved data and other information and to bring about a greater awareness of activities associated with managing and improving the environment of the Danube River Basin and, ultimately, the Black Sea, the following recommendations are made.

*****

2 RECOMMENDATIONS

2.1 It was noted that other existing relevant river commissions such as the International Commission for the Protection of the Rhine or International Commission for the Protection of the Elbe have many experiences with data and information exchange. It is recommended that these experiences should be taken into account when improving the data exchange facilities in the Danube countries.

2.2 It was agreed that there is a need to bring data from the TNMN together with other sources of data and information in order to provide a single point of enquiry that will allow all user requirements to be met. This need can be easily met by an Internet application (based on the world wide web (www)). Such an application is technically simple, relatively low cost, and would allow dissemination at a global level.

2.3 It is recommended that a demonstration www Home Page for the Danube River Basin be constructed so that the potential could be shown to interested parties in order to secure funding for a host organisation to manage and update the system.

2.4 It is recommended that only data and information that are officially approved for publication or public release should be placed on the system although ultimate responsibility for this would lie with owners of the information such as the database owners, individual countries and the Danube International Commission.

2.5 It is recommended that the Home Page should serve as a gateway or road map to existing data sources maintained by others at their own expense, therefore maximising data sources' use and minimising the costs to the host sources as well as avoiding duplication.

2.6 It was recognised that Environment Ministries and associated institutes of some countries of the Danube River Basin already had Home Pages on the www and it is recommended that these should form the basis for carrying data and information at the country level and be controlled and funded at that level.

2.7 It is recommended that other Danube countries should construct Home Pages for the purposes of disseminating information about their countries on a national and international basis and therefore benefit from greater public awareness and involvement in environmental issues.

2.8 It is recommended that non-governmental organisations (in the public and private sector) and research and educational institutes should be invited to participate in this process and to make their information available in a similar format.

2.9 It is recommended that linkages be developed within the Home Page with the information and communication systems of all relevant international agencies and activities.

2.10 It is recommended that linkages be developed with the European Environment Agency as mandated at the last meeting of Environmental Ministers in the Sofia Declaration in October 1995.

* * * * *

3 FUTURE ACTION:

3.1 Begin the consultation process with the Danube River Basin countries, in particular the Country Programme Co-ordinators, to gain their support in principle to these recommendations and to learn of their plans for national initiatives on the www.

3.2 In parallel, consult with the Danube Programme Co-ordination Units and the interim Danube International Commission to seek their views on the recommendations and level of support that could be provided.

3.3 Consult with the NATO Scientific Affairs Division to gain its support for follow-on funding for the production of the demonstration Danube Home Page facility.

3.4 Assuming satisfactory outcomes from 3.1 to 3.3, initiate a small team of experts from NATO member countries and Cooperation Partner countries to develop the demonstration facility on an existing web-server. This could be done through another, but smaller, ARW or through the NATO Expert Visits mechanism.

3.5 Inform all interested parties of the location of the demonstration facility and seek their views on its suitability to meet user needs and to make the necessary adjustments to the system (perhaps co-ordinated through a follow-on ARW).

3.6 Consult with possible hosts for the Home Page and achieve support for the maintenance and update costs from donor organisations.

List of Contributors

Mr. Robert Atkinson
Regional Environmental Center
1 Miklós tér
1035 Budapest, Hungary
T: 36 1 250-3401 F: 36 1 250-3403

Dr. Péter Bakonyi
VITUKI Consult Rt
Kvassay J. út 1
H-1095 Budapest, Hungary
T: 36 1 216-5811 F: 36 1 215-2245
email: vitukicons@attmail.com

Dr. Edward Bellinger
Central European University
Environmental Sciences & Policy
Nádor utca 9, Budapest, Hungary
T: 36 1 327-3021 F: 36 1 327-3031
email: herczegi@ceu.hu

Mr. Marc Braun
ICPR
Hohenzollernstr 18, Postfach 309
56003 Koblenz, Germany
T: 49 261 1 2495 F: 49 261 36572

Dr. Mitja Brilly
Univeristy of Ljubljana
Faculty of Civil Engineering
Ljubljana, Slovenia
T: 386 61 1254 333 F: 386 61 219897

Mr. Josef Drozd
Morava River Basin Authority
Drevarska 11, Brno 60175,
Czech Republic
T: 42 5 727 1321 F: 42 5 4121 1403

Ms. Marcela Fabianová
Slovak Hydrometeorological Institute
Jeseniova 17
83315 Bratislava, Slovakia
Fax: 42 7 374 195

Dr. Meta Gorišek
Min. Environment/Physical Planning
Vojkova 1a,
61000 Ljubljana, Slovenia
T:386 61 1784 533 F:386 61 1784 611

Dr. Jasna Grbovič
Hydrometeorological Institute of
Slovenia
Vojkova 1b
Llubljana, Slovenia

Mr. Wilhelm Kittinger
International Commission for the
Protection of the Danube River
Ministry of Agriculture and Forestry,
Section IV, Marxergasse 2
A-1030 Vienna Austria
T: 431 71 409 5012
F: 43171 409 5030

Dr. Tim Lack
WRc
Medmenham PO Box 16
Marlow Bucks SL7 2HD, UK
T: 44 1491 571531
F: 44 1491 579 094

Mr. Ferenc László
VITUKI Inst. of Water Quality Control
Kvassay Jenő út. 1
H-1095 Budapest, Hungary
T: 36 1 215 6140 F: 36 1 216 1514

Prof. Dr. Hans Liebscher
Federal Office of Water
Postfach 309
56003 Koblenz, Germany
T: 49 261 130 60 F: 49 261 130 6302

Dr. Irene L. Murphy
Colorado State University
2005 37th St. NW
Washington, DC 20007 USA
T: 1 202 337-2376 F: 1 202 342-6434
email: imurph@aol.com

Dr. H. Neiderländer
International Centre for Water Studies
Damrak 83
1012 LN Amsterdam, Netherlands
F: 31 20 6 22-83-84

Ron Orach
Metcalf and Eddy
30 Harvard Mill Square
Wakefield, MA 01880,USA
T: 1 617 246 5200 F: 1 617 245 0823

Dr. Atanas Paskalev
National Association on Water Quality
3 Oboristed St.
1504 Sofia, Bulgaria
T: 359 2 43 5129/ 87 42 51
F: 359 2 43 55 19

Mr. David Rodda
Consultant
Walton-on-Thames
Surrey, UK, KT 12 1EE
T and F: 44 1932 220 318

Mr. Karl Schwaiger
Ministry for Agriculture and Forestry,
Sec IV
Marxergasse 2
A-1030 Vienna Austria
T: 43 1 7140950-24 F: 43 1 714095-30

Sir Richard Southwood
Zoology Department University of
Oxford
South Parks Road, Oxford, UK
T: 44 1865 271 255 F: 44 1865 271248

Dr. Aurel Varduca
Research and Engineering Institute for
Environment
Spl Independentei 294 Sector 6
Bucharest 78, Romania
T: 40 1 637 30 60 F: 40 1 312 13 93

Mr. David J. Wangsness
USGS Water Resources Division
3039 Amwiler Road
Atlanta, Georgia 30360-2824 USA
T: 1 770 903-9100 F: 1 770 903-9199

List of Participants

Mr. Robert Atkinson
Regional Environmental Center
1 Miklós tér
1035 Budapest, Hungary
T: 36 1 250-3401 F: 36 1 250-3403

Dr. Péter Bakonyi
VITUKI Consult Rt
Kvassay Jenő út 1.
H-1095 Budapest, Hungary
T: 36 1 216-5811 F: 36 1 215-2245
email: vitukicons@attmail.com

Dr. Edward Bellinger
Central European University
Environmental Sciences and Policy
Nádor utca 9 Budapest, Hungary
T: 36 1 327-3021 F: 36 1 327-3031
email: herczegi@ceu.hu

Mr. Volodimir Bilikon
Ministry for Environmental Protection and Nuclear Safety
5 Khreshchatyk str
252601 Kiev, Ukraine
T: 380 44 2240411 F: 380 44 2298383

Mr. Marc Braun
ICPR
Hohenzollernstr 18, Postfach 309
56003 Koblenz, Germany
T: 49 261 1 2495 F: 49 261 36572

Mr. Josef Drozd
Morava River Basin Authority
Drevarska 11, Brno
60175, Czech Republic
T: 42 5 727 1321 F: 42 5 4121 1403

Ms. Marcela Fabianová
Slovak Hydrometeorological Institute
Jeseniova 17
83315 Bratislava, Slovakia
F: 42 7 374 195

Dr. Meta Gorišek
Ministry of Environment and Physical Planning
Vojkova 1a 61000 Ljubljana, Slovenia
T: 386 61 1784533
F:386 61 1784611

Mr. Peter Hlavatý
Ministry for the Environment
Hlboka 2
812 35 Bratislava, Slovakia
T: 42 7 492 451
F: 42 7 491 840

Dr. Jasna Grbovič
Hydrometeorological Institute of R Slovenia
Vojkova 1b
Llubljana, Slovenia

Mrs. Katerina Klimenko
Ministry of Environmental Protection
5 Khreschchatyk St.
252001 Kiev, Ukraine
T 380 44 228 0644 F: 380 44 229 7430

Mr. Jan Kubat
Czech Hydrometeorological Institute
Na Sabatce 17, Praha 4
T: 42 2 401 0800
F: 42 2 401 0800

Dr. Tim Lack
WRc
Medmenham PO Box 16
Marlow Bucks SL7 2HD, UK
T: 44 1 491 571 531
F: 44 1 491 579 094

Prof. Dr. Hans Liebscher
Federal Office of Water
Postfach 309
56003 Koblenz, Germany
T: 49 261 130 60 F: 49 261 130 6302

Mr. Ivan Milushev
Ministry of Environment
67 West Gladstone St.
Sofia, Bulgaria
T: 359 2 955 9421 F: 359 2 52 16 34

Dr. Irene L. Murphy
Colorado State University
2005 37th St. NW
Washington, DC 20007 USA
T: 1 202 337-2376 F: 1 202 342-6434
email: imurph@aol.com

Dr. Corneliu A. L. Negulescu
Research and Engineering Institute for Environment
Spl. Independentei 294 Sector 6
Bucharest 78, Romania
T: 40 1 637 30 60 F: 40 1 312 13 93

Dr. H. Neiderländer
International Centre for Water Studies
Damrak 83
1012 LN Amsterdam
Netherlands
T: 31 20 627 53 87
F: 31 20 622 83 84

Mr. Željko Ostojič
Deprt. for International Cooperation
Ulica grada Vukovara 78
HR-10000 Zagreb, Croatia
T: 385 1 6110-522 F: 385 1 510-137

Dr. Atanas Paskalev
Bulgarian National Association on Water Quality
3 Oboriste St. 1504 Sofia, Bulgaria
T:3592 435129/874251F:359 2 435519

Mrs. Eszter Poroszlai
Min Environment Protection/Regional Policy
Fő u. 44-50, 1011 Budapest, Hungary
T: 36 1 201-3843 F: 36 1 201-3735

Mr. David Rodda
Consultant
1, The Chestnuts
Walton-on-Thames
Surrey, UK, KT 12 1EE
T/F: 44 1932 220 318

Mr. Karl Schwaiger
Ministry for Agriculture and Forestry, Sec IV
Marxergasse 2
A-1030 Vienna Austria
T: 43 1 7140950-24 F: 43 1 714095-30

Sir Richard Southwood
Zoology Department University of Oxford
South Parks Road
Oxford, UK
T: 44 1865 271 255
F: 44 1865 271248

Dr. Aurel Varduca
Research and Engineering Institute for Environment
Spl. Independentei 294 Sector 6
Bucharest 78, Romania
40 1 637 30 60 F40 1 312 13 93

Mr. David J. Wangsness
USGS WR Div 3039 Amwiler Road.
Atlanta, GA 30360 USA
T: 1 770 903-9100 F: 1 770 903-9199

Abbreviations and Acronyms

AEWS	Accident Emergency Warning System
AOX	Absorbable halogenated organics
BAT	Best Available Techniques
BEP	Best Environmental Practice
BOD	Biological Oxygen Demand
BOT	Build-Operate-Transfer
CEE	Central and Eastern Europe
CEEC	Central and Eastern European countries
CIS	Commonwealth of Independent States
COD	Chemical Oxygen Demand
CPC	Country Programme Coordinator
DEFF	Danube Programme Data Exchange File Format
DO	Dissolved oxygen
DRPC	Danube River Protection Convention
EAP	Environmental Action Plan
EBRD	European Bank for Reconstruction and Development
EC	European Commission
ECE	UN Economic Commission for Europe
ECU	European Currency Unit
EIB	European Investment Bank
EPA	Environmental Protection Agency
EPDRB	Environmental Programme for the Danube River Basin
EU	European Union
GEF	Global Environment Facility
GIS	Geographical Information System
HCB	Hexa-chloro-benzene
IBRD	International Bank for Reconstruction and Development (The World Bank)
ICPDR	International Commission for the Protection of the Danube River
ICPR	International Commission for the Protection of the Rhine
ISO	International Standards Organisation
IUCN	World Conservation Union
m	metre
m^3/s	cubic metres per second
km	kilometre
MECU	Million European Currency Units
MLIM	Danube Programme Monitoring, Laboratory, Information Management Sub-Group
NAP	National Action Programme
NAS	National Academy of Sciences (United States)

NATO	North Atlantic Treaty Organization
NGO	Non-Governmental Organisation
NRA	National Rivers Authority (United Kingdom)
OECD	Organisation for Economic Co-operation and Development
PCU	(Danube) Programme Coordination Unit
PHARE	EU Prograrnme of assistance for economies of central and eastern Europe
RAP	Rhine Action Programme
REC	Regional Environmental Center, Budapest
sq km	square kilometre
TNMN	Danube Programme Trans-National Monitoring Network
TVS	Total volatile solids
UNDP	United Nations Development Programme
UNEP	United Nations Environment Programme
USAID	United States Agency for International Development
USEPA	United States Agency for Environmental Protection
USGS	United States Geological Survey
VITUKI	Hungarian Water Resources Research Institute
WWF	World Wide Fund for Nature
WWTP	Waste water treatment plant(s)

Index

MIX
Papier aus verantwortungsvollen Quellen
Paper from responsible sources
FSC® C105338

If you have any concerns about our products,
you can contact us on
ProductSafety@springernature.com

In case Publisher is established outside the EU,
the EU authorized representative is:
Springer Nature Customer Service Center GmbH
Europaplatz 3, 69115 Heidelberg, Germany

Printed by Libri Plureos GmbH
in Hamburg, Germany